Building Research Establishment Report

The building limestones of the British Isles

Elaine Leary, BSc
Building Research Station

Department of the Environment
Building Research Establishment

London, Her Majesty's Stationery Office

Full details of all new BRE publications
are published quarterly. Requests for placing
your name on the BRE mailing list should be
addressed to: Mailing List (Commands)
Building Research Establishment
Garston, Watford, WD2 7JR

Cover photograph
Part of Prior Park College built by
Ralph Allen in Combe Down Stone (see text p16)

ISBN 0 11 671365 8
©Crown copyright 1983
First published 1983

Limited extracts from the text may be reproduced
provided the source is acknowledged. For more
extensive reproduction, please write to the Publications
Officer at the Building Research Establishment.

Contents

	Page
Foreword	v
Introduction	1
Chapter 1 Geology	3
Chapter 2 Selection, durability and the interpretation of the test results	6
Indirect tests	6
Porosity	6
Saturation coefficient	6
Porosimetry by water suction or mercury intrusion	7
Capillarity	7
Combinations of pore parameters	7
Direct tests	7
Frost test	7
Crystallisation test	7
Conclusion	8
Durability of carboniferous limestones	8
Notes on the presentation of test results	8
Chapter 3 The quarries and their stones	11
Ancaster Stone	12
Ballinasloe Stone	14
Bath Stone	
Combe Down – Mount Pleasant Quarry	15
Combe Down – Upper Lawns Quarry	16
Monks Park	17
Stoke Ground Base Bed	19
Westwood Ground	20
Clipsham Stone	22
Big Pits and Longdales Quarries	23
Medwells Quarry	25
Daglingworth Stone	26
Doulting Stone	27
Fenacre Stone	29
Guiting Stone	
Coscombe Quarry	30
Cotswold Hill Quarry	32
Ham Hill Stone	33
Happylands Stone	35
Hopton Wood Stone	36
Hornton Stone	37
Hovingham Stone	38
Kentish Ragstone	39
Ketton Stone	41
Kilkenny Marble	43
Lecarrow Stone	44
Mandale Stone	45
Nash Rocks	46
Navan Stone	47
Orton Scar Stone	48
Penmon Stone	49

(continued)

	Page
Portland Stone	50
Bowyers Quarry	52
Independence Quarry	53
Kingston Minerals' Quarries	54
Coombefield Quarry	
Fancy Beach Quarry	
Weston Quarry	
Sheat Quarry	57
Purbeck Stone	59
Purbeck Limestone	
California Quarry	61
Downs Quarry	63
Keates Quarry	64
Landers Quarry	65
Swanage Quarry	66
Purbeck-Portland Stone and Purbeck Marble	
Haysom's Quarries	68
Salterwath Stone	70
Stowey Stone	71
Swaledale Fossil Stone	72
Taynton Stone	73
Totternhoe Stone	74
Ulverston Marble	76
Weldon Stone	77
Wroxton Stone	79

Appendix A Description of the tests — 80

Velocity of sound — 80
Porosity — 80
Saturation coefficient — 80
Microporosity — 80
Capillarity — 81
  Description of Belgian and French capillarity coefficients — 81
    Belgian capillarity coefficient, G — 81
    Belgian capillarity coefficient, GC — 81
    French capillarity coefficient, A — 82
Crystallisation test — 82

Appendix B Appraisal of European test procedures — 83

Velocity of sound — 83
Capillarity test — 83
  Belgian capillarity coefficient, G — 83
  Belgian capillarity coefficient, GC — 84
  French capillarity coefficient, A — 84
Conclusion — 84

Acknowledgements — 86

References — 87

Bibliography — 87

Index — 88

Colour plates of Green College Oxford, a quarry, petrological thin sections, and stones — Inset

Foreword

Although stone is a traditional building material in the British Isles, most of the classic books on stone have long been out of print, and there is consequently little published information to consult. British Standard Codes of Practice deal with design and repair, but give little guidance on the choice of a suitable stone. To the newcomer, the wide variety of available stone can be bewildering.

This book sets out to fill the gap, by describing the British limestones and listing buildings where they can be seen. It takes the bold step of indicating the comparative durability of each stone, and links this with the environment and parts of a building where the stone may be used safely.

The stone industry has collaborated fully with the Building Research Establishment in the preparation of this book. It is hoped that their co-operation will be rewarded by the increased use of the right stone in the right place.

I Dunstan
Director, Building Research Establishment, 1983

Introduction

Limestone has provided the fabric for some of the most magnificent buildings in the southern part of Britain – great cathedrals such as Wells, St Paul's and Ely, and houses such as Audley End and Montacute. Whole regions like the Cotswolds, and cities like Bath, Oxford and Cambridge, owe most of their distinctive character to the limestone of their buildings. Stone has proved its worth over the centuries. Architects are now turning to limestone again for new buildings where it can provide structural walls or a durable and decorative cladding (see, for example, Plate 1).

This book was compiled to provide designers and architects of new buildings, and others involved in restoration work, with a guide to the currently available limestones of the British Isles. The aim of the study on which this book is based was to visit, and test the stone from, only those quarries providing dimension (ashlar) stone. Subsequently one or two of the quarries were found to provide random walling only, but they have been included in the book to give a complete record of all the stone tested. Magnesian limestones are not included.

Chapter 1, on geology, describes the way in which limestones were formed, the terms associated with quarrying, and the design considerations in the use of limestones in buildings. Chapter 2 discusses selection and durability of limestones and assessment of the test results.

The bulk of the book (Chapter 3) describes the quarries visited and their stones. Visits started in 1980. Several reference books were used in an effort to find all the quarries which were actively extracting stone for building use. The survey of quarries was carried out in conjunction with the Department of the Environment's Directorate of Ancient Monuments and Historic Buildings. The Directorate sent each quarry a letter asking for their cooperation in carrying out the survey. Staff from either the Directorate or the Building Research Station (BRS) visited the quarries to collect information and representative stone samples for testing. Press notices appeared in several journals, notably *Stone Industries, Construction News* and *Building Trades Journal,* inviting active quarries with whom we had not been in touch to contact BRS.

The following information on each quarry is given:

Exact location together with an Ordnance Survey grid reference for the quarries in Great Britain

Size of the quarry and an indication of the potential stone reserves for future exploitation

Geological details

Description of the stone and the sizes available

Description of the applications for which the stone has been used and reference buildings where it can be seen

Results of the testing programme

Durability class.

The inset of coloured photographs includes pictures from the main beds of most of the stones.

The sets of reference buildings were compiled largely from information provided by each quarry and by reference to many of the books listed in the Bibliography. The Stone Federation was also consulted. Every effort was made to give each set as wide a geographical spread as possible. One of the best places to see a particular stone may be the local church or local housing. However, it would be profitable for any reader interested in a particular stone to contact the relevant quarry for a list of their latest contracts. Similarly, as quarries are constantly opening, or closing, readers should always consult the *Natural stone directory*[1] for up-to-date information on most of the active quarries. Eight quarries have opened since the survey visits were completed; their addresses are given but they opened too late to be included in the full testing programme.

At the end of the visits so many samples had been collected that it seemed opportune to carry out some additional European tests as well as the customary British ones. The aim was to discover if the durability indicated by the European tests gave a realistic prediction of a stone's performance in the climate of the British Isles. Thus, the tests themselves were 'put to the test' as well as the stone samples. The tests are described in Appendix A whilst Appendix B discusses the European tests. The tests which were carried out are:

Measurement of the velocity of sound[2]
Measurement of porosity and saturation coefficient[3]
Measurement of microporosity[3]
Capillarity test[4]
Crystallisation test[2,3].

Chapter 1 Geology

Geologically, limestones are classed as sedimentary rocks and essentially they consist of calcium carbonate. The limestones described in this book were laid down in the following four geological ages:

	Age in millions of years
Cretaceous	65 to 136
Jurassic	136 to 190 – 195
Carboniferous	280 to 345
Silurian	395 to 430 – 440

The most important building limestones are of Jurassic age.

The British limestones run broadly in a belt from Dorset, through the Cotswolds and Oxfordshire and up into Northamptonshire and Lincolnshire, to just north of the Humber. Some of the older stones are found in Cumbria and Wales. Locations of the quarries are shown in Figure 3 at the beginning of Chapter 3, and a typical quarry is shown in Plate 2.

In most of the limestones with which we are concerned the calcium carbonate was deposited either as loose particles from the skeleton and shells of once living organisms, or chemically. The oolitic limestones are an important example of a chemical deposit. They consist mainly of small rounded grains of calcium carbonate known as ooliths, each of which was deposited in concentric layers around a sand grain or piece of shell which acted as a nucleus. The name derives from the appearance of the stone in that the small grains or ooliths resemble fish roe (Greek: *oion* – an egg, *lithos* – a stone).

The calcium carbonate was slightly dissolved by the action of acidic water containing carbon dioxide. This dissolved calcium carbonate was then redeposited amongst the loose particles in patches, cementing them together and ultimately producing the beds of stone. All this deposition took place at the bottom of seas and lakes.

The stone was then compacted by the weight of further sediments above and sometimes by pressures caused by movements of the earth's surface. Thus older limestones tend to be more crystalline and dense since they have been subjected to greater pressures.

The structure of oolitic limestone varies considerably. Ketton Stone is regarded as the classic oolitic stone in that it seems to consist almost entirely of the small spherical grains with very little cementing matrix. In other stones, eg Bath Stone, the ooliths of calcium carbonate seem to be set in a continuous matrix of calcium carbonate. Some oolitic stones contain numerous shell fragments whilst others, like Monks Park, have a fine, even texture. It is the oolitic limestones which have provided the most famous of the building limestones, owing to their workability.

Generally, the more coarse a stone is and the more shell fragments it contains, the more durable it will be. A really shelly stone like Portland roach is very durable. In addition, several limestones, in particular the hard Carboniferous stones, take an excellent polish. Many different colours of limestone are available ranging from black, grey and white through to buff, yellow and orange. Blue is another common colour. The colour is due to the presence of other minerals – often oxides of iron.

Sedimentary rocks are laid down in layers. One layer of particles drifts to the bottom of the sea or lake to lie on the top of the previous layer. Thus these layers, which are known as beds, are laid down in parallel on top of one another. Each bed is bounded top and bottom by a bedding plane which indicates where interruptions occurred in the laying down of the sediment. The depth of each bed is measured from one bedding plane to the next. The bedding planes are very important when the stone is quarried since they indicate where one bed can be separated, and lifted off, from the bed below.

Joints are also very important from the point of view of quarrying. As each layer of a sedimentary rock dries out, shrinkage cracks or joints appear. The run of each joint is limited by the bedding plane. Strongly developed joints which do cross bedding planes are known as master joints. Joints frequently occur at right angles to each other and well jointed rocks are more easily extracted; blocks of stone of bed height can be lifted off as delimited by the bedding planes and vertical joints.

When the rock is not well jointed or where the joints are too far apart to permit the extraction of blocks of manageable size, then the stone must be split. This is typically done with 'plug and feathers'. Firstly, a row of holes, anything from 230 to 450 mm apart, is drilled along the line where the block is required to split. Two long strips of metal ('feathers') with a long wedge ('plug') between them are inserted into each hole. Then the plugs are hammered home down between the feathers. The plugs are hammered a little at a time in series so that the block is submitted to an even pressure and splits cleanly.

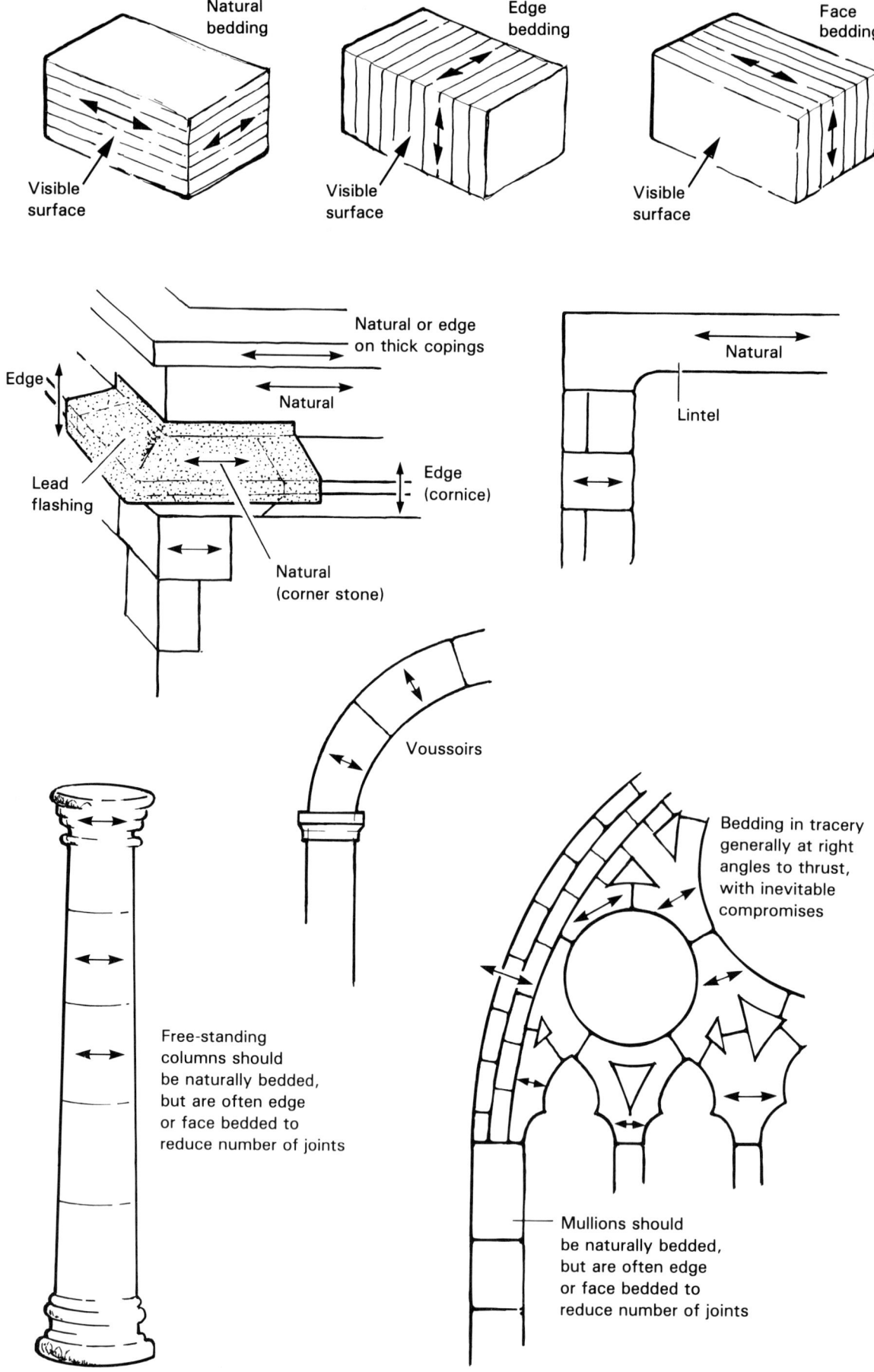

Figure 1 Illustration of bedding principles. The double-headed arrows indicate the bedding of the stone as placed in the building. For some features, such as corner projections and gargoyles, no orientation of the bedding is entirely satisfactory — the tops of these are best covered with lead. Lead flashing over a cornice is always to be recommended

The face of the quarry is the depth of the stone exposed during working and is usually determined by purely economic reasons.

The layered structure of sedimentary rocks must be borne in mind when placing stone in buildings. A bed of stone is rather like a ream of paper. Both must be restrained to prevent each layer or page separating and being lost. Thus stone should be placed on its natural bed, ie so that each layer is horizontal, just as they were laid down geologically. Edge-bedding, where the layers are vertical and perpendicular to the face of the wall, is used for cornices and copings (but not for the corner stones, which should be naturally bedded). The voussoir stones of arches are laid with the bed at right angles to the thrust of the arch. Note that, although a freestone is defined as a stone which can be sawn or worked in any direction, these general bedding principles, illustrated in Figure 1, still apply.

One outcome of placing stone on bed is that there is a limit to the depth of stone available. Therefore, when specifying the course heights, designers need to know the depth of stone which can be obtained from each bed. This is known as 'the depth of stone on bed'. It is not necessarily the same as the total depth of the bed since there may be some waste within each bed which must be discarded. Equally, there may be minor bedding planes within a single major bed, so that the largest block that can be extracted may be much smaller than the total depth of the bed. Before deciding to use a particular stone, architects should visit the quarry to find out the depth of stone on bed and to discuss the likely quantities involved. There is no point in specifying a course height of 450 mm if the quarry can only supply 300 mm on bed, and not every quarry can supply large quantities of stone.

Quarries can be dangerous places and some do not work every day, so visitors are advised not to arrive unannounced.

Chapter 2 Selection, durability and the interpretation of the test results

Having decided to specify stone in a building, which stone should one choose? Perhaps the easiest decision to make first is on the basis of colour and texture — will the stone fit in with its surroundings or provide a contrast as required? However, bear in mind that some stones look different on wetting and can, after a few years, weather to a completely different colour. Therefore, one should look at buildings completed some years ago to see the colour to which the stone weathers. Looking at old buildings also gives some idea of the durability of the stone, although of course the current output may differ. One can judge how well the stone has performed by looking for spalls, blisters, friable surfaces and any crumbling or splitting of the arrises. A rough texture is normal for some stones after weathering and is no indication of poor durability.

The limestones of the British Isles offer a wide range of stone in terms of colour and texture. Unfortunately, they also offer a wide range of quality in terms of durability. The quality of the output of the quarry can change and so the reputation of a stone is not necessarily a reliable guide to the quality of the current output. However, there are very few stones which will not give satisfactory service in sheltered situations for the 60-year design life often taken for domestic buildings. Durability will need to be assessed only when the stone is intended for longer life, for harsher exposures (eg string courses or copings) or for more severe environments.

But what determines durability? It is not chemical composition, since all limestones are virtually identical in this respect; what is more important is the physical structure of the stone. Although a stone may look solid on first sight, they all contain numerous minute pores. The term pore here denotes the small spaces both within and between the particles of calcium carbonate. It is the pores that take up water on wetting and it is also here that any harmful products, resulting from the action of pollution, accumulate.

Plate 3 shows petrological thin sections of two oolitic limestones whose pores have been filled with a coloured resin. Petrology is the study of the origin of rocks and these thin sections are used to determine mineral composition and structure. It has been known for a long time that a stone with a high proportion of very small pores is less durable than a stone that has larger pores. Thus the stone in Plate 3(a) will be less durable than that in Plate 3(b). However, the interpretation of these sections is not always so obvious and a great deal of experience is called for. Since the sections are expensive to produce, cheaper but less direct methods of characterising pore structure are used.

Unfortunately, none of these methods are suitable for use at the quarry. At present, the only practical test which a quarry man can carry out is to strike a block of stone with an iron bar. If it rings true then it contains no large internal flaws which would be revealed on sawing.

Strength tests are seldom necessary since the weakest stone can be expected to withstand the loads imposed upon it in normal usage. After all, the stone is subjected to huge loads as it lies in the quarry face. However, any stone will fail through undue stresses caused by uneven settlement, insufficient allowances for thermal movement or the pressures from corroding iron cramps.

The tests carried out for this survey are described in Appendix A. They fall into two categories:

1 Indirect tests of durability based on assessment of pore structure
2 Direct tests of durability which subject the stone to the type of conditions it will meet in a building, although in a more aggressive form.

INDIRECT TESTS
Porosity
Porosity is the volume of a stone's pore space expressed as a percentage of the stone's total volume. It can range from 1 to 2 per cent for low-porosity stones up to as much as 40 per cent; 10 to 25 per cent are typical values for the more porous stones. Porosity alone is not a reliable guide to durability since it is the size of the pores that is critical and the porosity gives no indication of pore sizes.

Saturation coefficient
The saturation coefficient is a measure of the extent to which the pores become filled when a stone is allowed to absorb water for a standard time. It is a characteristic property of the stone and depends on the size of the pores and not on the porosity. Together with porosity it gives some idea of the amount of water that the stone will absorb under natural exposure conditions.

Values can range from 0.40 to 0.95. A high value indicates a stone with a high proportion of small pores. Unless its porosity is very low such a stone is likely to be damaged by a sharp frost occurring after a shower of rain or by the effects of pollution. Low values of the coefficient indicate durable stones. Unfortunately, many stones fall into the middle area of 0.66 to 0.77 and in this region the saturation coefficient on its own is an unreliable guide to durability.

It should also be noted that measurement of this coefficient for low-porosity stones (ie porosity less than 5 per cent) is rarely useful because these stones are usually very durable even if the coefficient is high.

Porosimetry by water suction or mercury intrusion
A variety of techniques are available for detailed investigation of the range of pore sizes in stone. The two that are most widely used are the suction plate technique and mercury porosimetry. They are based on the fact that the pressure required to suck water out of a full pore or force mercury into an empty pore depends on the size of the pore. In the former case, the test piece is subjected to a range of negative pressures or suctions until it will give up no more water at each pressure. In the latter, the mercury content at various pressures is measured. The results can then be used to derive the range of pore sizes.

With the suction plate technique those pores with an effective diameter of less than 5 micrometres (μm) are termed micropores and the proportion of the total pore space that is accounted for by these pores is termed the microporosity. It is not necessary to measure the entire pore size distribution in order to obtain the microporosity; the suction at which it is possible to extract water from those pores with an effective diameter greater than 5 μm can be calculated experimentally.

Stones with a low microporosity — less than 30 per cent — are durable; stones with a high microporosity — greater than 90 per cent — are not durable. Again there is a middle area in the 50 to 80 per cent region where, on its own, the value of microporosity is an unreliable guide to durability. Further, the measurement of this parameter for low-porosity stones (ie porosity less than 5 per cent) is rarely useful because these stones are usually very durable even if microporosity is high.

In mercury porosimetry an arbitrary line is drawn at the point where 10 per cent of the pore space has been filled with mercury (this is broadly equivalent to the point where 90 per cent of the pore space would be filled with water in the water suction method). The pore diameter corresponding to that point is termed 'd_{10}' and is used in Belgian selection procedures[2] in which a stone with a d_{10} greater than 2.5 μm and a saturation coefficient less than 0.80 is considered to be frost resistant.

The mercury porosimetry technique was not used during the test procedures for this survey. The applicability of its criteria to the climate of the British Isles is yet to be established. But it is worth noting that there is a broad relationship between d_{10} and microporosity — the smaller the value of d_{10}, the higher the microporosity.

Capillarity
The capillarity test monitors the uptake of water by initially dry specimens in a shallow tray of water. The results are used by the French to derive a capillarity coefficient. Stone is then specified for use in different zones of a building on the basis of its capillarity coefficient and other parameters[2]. The purpose is to ensure, for instance, that stone is not used as a plinth course if it absorbs water readily. However, the coefficient does not appear to be entirely satisfactory. For example, it would bar the use of Portland Stone in plinth courses whereas Portland is very successfully used for plinths in the British Isles.

The Belgians also use capillarity test data and they derive two coefficients which they propose as measures of durability. However, evaluation of these coefficients during the test programme for this survey has shown that the results give no better indication of durability than the saturation coefficient[5] (see also Appendix B for discussion).

Combinations of pore parameters
A better indication of durability may sometimes be obtained by combining two parameters. Honeyborne and Harris[6] showed that durability assessment was enhanced if the porosity and saturation coefficient were considered together. The French use a combination of porosity and saturation coefficient as an indirect frost test to assess the suitability of stone for use in four exposure zones of buildings[4]. The combined use of microporosity and saturation coefficient has been shown to be a useful guide to the durability of Portland Stone[7].

DIRECT TESTS
Frost test
Freeze/thaw tests are used in French test procedures but were discontinued in the British Isles some years ago because of difficulties in establishing a realistic laboratory test on a small scale. A natural frost test, where specimens are left out in winter in trays of water, is more reliable but can take years to complete. Frost tests were not used for the present survey.

Crystallisation test
The main cause of stone decay in the British Isles is the crystallisation of soluble salts within the pores of stone[3]. These salts are derived from several sources ranging from rising damp to the acid gases of atmospheric pollution causing changes to the chemical constituents of the stone. The test which best amplifies the effects of pollution is the crystallisation test because it too causes soluble salts to crystallise in the pores of the stone.

Samples of stone are submitted to cycles of immersion in a solution of sodium sulphate followed by drying in a humid oven. The greater the weight loss is, the less durable the stone. The crystallisation test is a comparative one and reference samples of stone of known durability are included against which the unknown can be assessed. The limestones are subdivided into six durability classes, A to F, which are defined in Table 1. The classes are defined by the exposure zones of a building (set out in Figure 2) in which a stone can safely be used. Ideally, the reference samples should be of stone that has been used for the same building application and in the same part of the country as that proposed for the unknown. This is rarely possible but Table 1 provides a basis for adjusting from one environment to another.

Assume, for example, that the stone under test is intended for use in Zone 3 in an unpolluted, frost-free, coastal area. Table 1 shows that it will need to be at least Class C and preferably Class B. Reference stones are available whose performance is known only in a highly polluted, frost-free inland area. One of them (stone 1, say) gives good service in Zone 2 but is less good in Zone 1. Table 1 would thus rank stone 1 as Class B. The other stone (stone 2) gives adequate service in Zone 4 but is not suitable for Zone 3 and is thus given Class E. In the crystallisation test, the unknown stone will need to perform nearly as well as stone 1, and very much better than stone 2.

The inclusion of reference samples each time the crystallisation test is run is necessary because any slight variation in the experimental procedure can produce quite large differences in the absolute values obtained but not in the relative values. The other drawback of this test is that it takes several weeks to run. But, despite this, it is still the most reliable.

CONCLUSION

At the present time, the crystallisation test is regarded as the best single test for assessing the durability of an unknown porous limestone. For particular stone types other tests may be better since they are cheaper and quicker. For example, Bath stones with a saturation coefficient of less than 0.60 and Portland Stone with a microporosity less than 30 per cent are thought to be good durable stones.

DURABILITY OF CARBONIFEROUS LIMESTONES

The results for the older Carboniferous Limestones show that the durability of these stones is excellent – Class A. This means that they can confidently be used in the harshest exposures. However, one rarely sees cut and dressed Carboniferous Limestone in a building because it is so hard to work and often the depth of stone on bed is small. One must turn to the freestones of the Jurassic period for the more workable stones.

NOTES ON THE PRESENTATION OF TEST RESULTS

Each limestone in this book has been allocated a durability class on the basis of the crystallisation tests carried out on the samples collected for the survey. However, if this class is substantially different from what we would expect based on our past experience, we have commented on this immediately after the table of results for each stone. The classes are defined in Table 1 and the zones of use are illustrated in Figure 2. Thus, Table 1 shows where in a building and under which climatic conditions a stone may confidently be used.

All of the tests, except for the microporosity measurement, were carried out on samples of varying lengths with a square cross-section of 40 mm side. Two samples from each bed were used and, where possible, one was cut parallel to bed and the other perpendicular to bed. For the velocity of sound test the samples were 150 mm long. After this test the samples were cut down to 100 mm for the porosity, saturation coefficient and capillarity tests. These same pieces were then cut to produce two cubes of 40 mm side for the crystallisation test. Thus, four cubes from each bed were put into the crystallisation test. The measurement of microporosity was carried out on small pieces which were 15 × 15 × 5 mm.

In the table of results for each stone, some figures for the velocity of sound test and the capillarity test are given in parentheses. These two tests require samples which must be cut with a particular orientation with respect to bed. In the former test, samples must be cut parallel to bed, whilst the latter test requires samples which are cut perpendicular to bed. It was not possible to cut both types of sample for every stone because some samples provided were either too small or too shallow-bedded, so results for those samples which were not of the correct orientation are given in parentheses. They have been included since, in many cases, there is little variation between each type of sample.

Table 1 Effect of environment on the suitability of limestones for the four exposure zones of a building

	Suitability zones for various limestones in a range of climatic conditions							
	Inland				Exposed coastal			
Limestone durability class	Low pollution		High pollution		Low pollution		High pollution	
	No frost	Frost	No frost	Frost	No frost	Frost	No frost	Frost
A	Zones 1-4	Zones 1-4	Zones 1-4	Zones 1-4	Zones 1-4	Zones 1-4	Zones 1-4	Zones 1-4
B	Zones 2-4	Zones 2-4	Zones 2-4	Zones 2-4	Zones 2-4	Zones 2-4	Zones 2*-4	Zones 2*-4
C	Zones 2-4	Zones 2-4	Zones 3-4	Zones 3-4	Zones 3*-4	Zone 4	—	—
D	Zones 3-4	Zone 4	Zones 3-4	Zone 4	—	—	—	—
E	Zone 4	Zone 4	Zone 4*	—	—	—	—	—
F	Zone 4	Zone 4	—	—	—	—	—	—

*Probably limited to 50 years' life

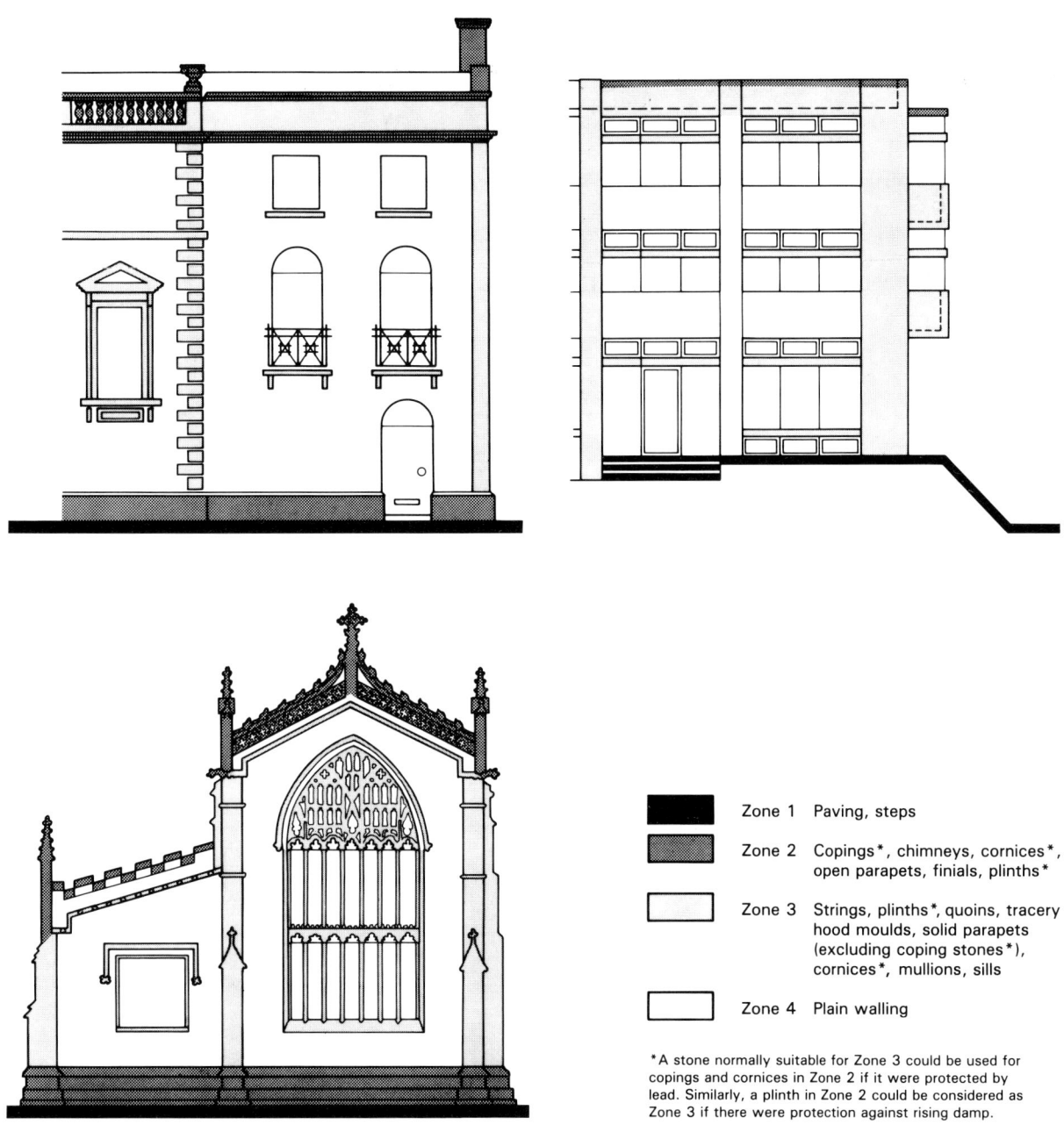

Zone 1 Paving, steps

Zone 2 Copings*, chimneys, cornices*, open parapets, finials, plinths*

Zone 3 Strings, plinths*, quoins, tracery hood moulds, solid parapets (excluding coping stones*), cornices*, mullions, sills

Zone 4 Plain walling

*A stone normally suitable for Zone 3 could be used for copings and cornices in Zone 2 if it were protected by lead. Similarly, a plinth in Zone 2 could be considered as Zone 3 if there were protection against rising damp.

Figure 2 Exposure zones of a building in which stone can be used

Figure 3 Locations of the limestone quarries and mines of the British Isles (opposite)

1 Portland (6 quarries)	10 Monks Park	19 Wroxton	28 Swaledale Fossil
2 Purbeck (6 quarries)	11 Kentish Ragstone	20 Hornton	29 Ulverston Marble
3 Fenacre	12 Daglingworth	21 Weldon	30 Orton Scar
4 Ham Hill	13 Taynton	22 Ketton	31 Salterwath
5 Doulting	14 Totternhoe	23 Clipsham (3 quarries)	32 Penmon
6 Stowey	15 Guiting – Coscombe Quarry	24 Ancaster	33 Kilkenny
7 Combe Down (2 quarries)	16 Guiting – Cotswold Hill Quarry	25 Hopton Wood	34 Ballinasloe
8 Hayes Wood (Stoke Ground Base Bed)	17 Happylands	26 Mandale	35 Lecarrow
	18 Nash Rocks	27 Hovingham	36 Navan
9 Westwood Ground			

Chapter 3 The quarries and their stones

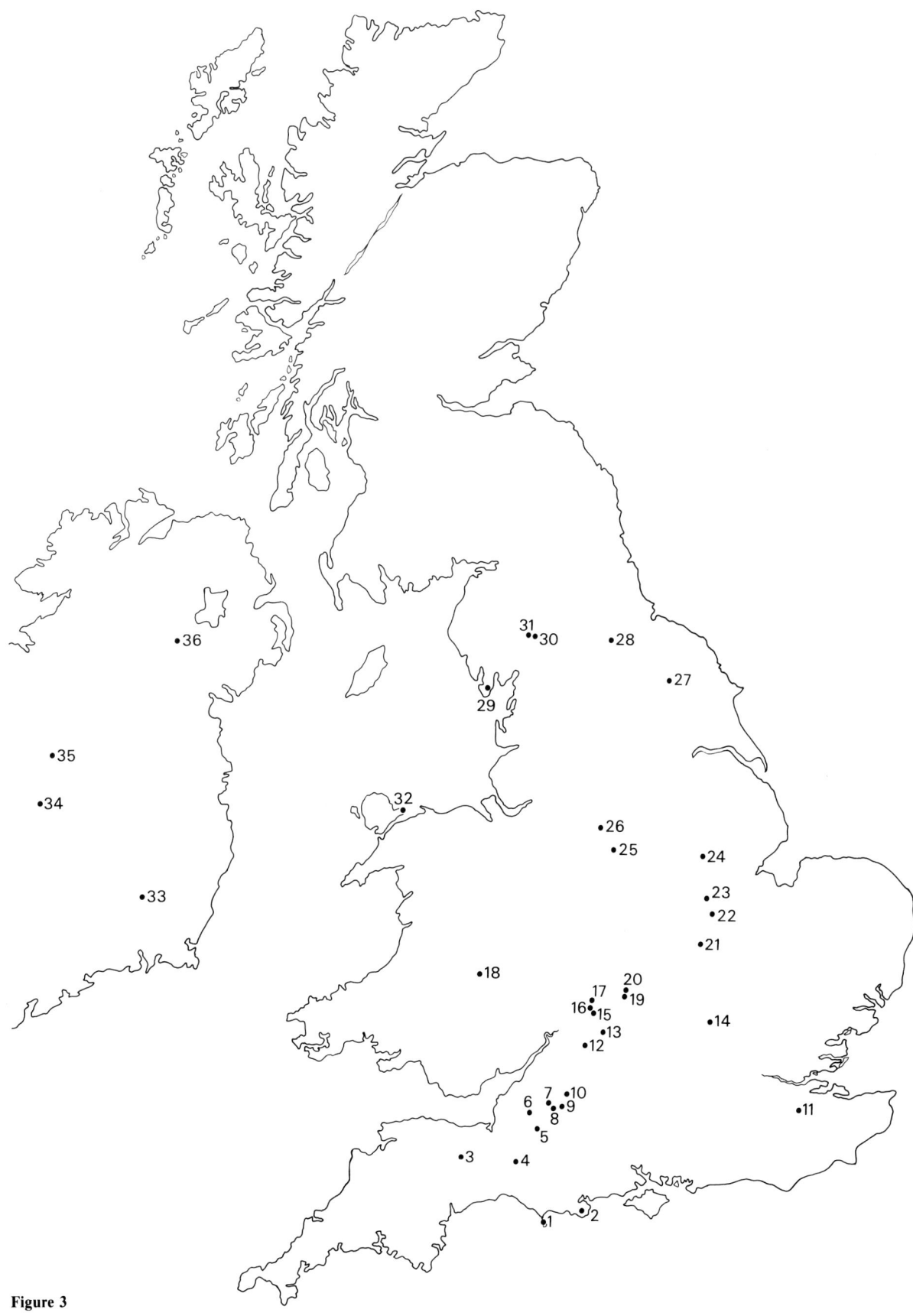

Figure 3

ANCASTER STONE

Grid reference SK 991 407

Owners The Gregory Quarries Ltd, 184 Nottingham Road, Mansfield, Notts.
Tel: Mansfield (0623) 23092

Quarry location Ancaster, Nr Grantham, Lincolnshire

The quarry is south of the village of Ancaster off the B6403. Approaching the village from the south on the B6403 you turn right towards Wilsford by the Barkston Heath Airfield. At the first crossroads turn right onto the old Roman road. There is then a wood on the left and shortly after this an unmarked track on the left just before farm buildings which are on the right. Follow this track, bearing left at the trees to reach the quarry which is some distance from the road.

The stone was worked by both the Romans and the Saxons. The present quarry, which is surrounded by old workings, has been worked since 1957 (approximately) and there is enough stone for a further 20 years' work. In addition the quarry can expand to an adjacent field.

Petrography *(Plate 4)*

Ancaster Stone is an oolitic limestone from the Lincolnshire Limestone formation of middle Jurassic age. There are three beds of stone under 7.8 m of overburden:

Weather bed — a warm-brown, shelly stone. The total depth of stone in this bed is 2.1 to 2.4 m and the depth of stone on bed available as individual blocks is about 530 mm. The stone from the bottom of the bed is very shelly and takes an excellent polish.

Hard white — a creamy coloured stone of uniform texture with very little shell. The depth of this bed is 2.7 to 3.0 m and the depth of individual blocks of stone on bed is 450 to 500 mm.

Freestone — a buff coloured stone which is easier to work than the hard white and contains a few shell fragments. The depth of this bed is 1.5 m and the depth of individual blocks on bed is about 530 mm.

The beds of stone tend to slope to the extent that the freestone bed ran out altogether some years ago. However, as the face of the quarry was worked back the freestone reappeared and is now available once more.

Reference buildings

The hard white bed of stone is today the most popular of the three stones available at Ancaster. It is used for ashlar work and for carving. The freestone can also be used for ashlar provided it is sawn and not cleaved with a hammer since it is apt to split along sand veins. When it does split it is used for random walling.

The weather bed is used for flooring and fireplaces and was used in the 1960s for paving the South Transept Chapel in Ely Cathedral. Polished weather bed was used for the reception area of the Royal Commonwealth Society in London in approximately 1958–59. In the present quarry there is only one weather bed — the brown. The red weather bed referred to in some older books is not present. However, the present weather bed can contain substantial amounts of pink and blue stone. The latter is similar to blue Clipsham but tends to be more shelly.

In the fifteenth century Ancaster was used in all the principal churches of Norwich including St Peter Mancroft. At Tattershall Castle in south-east Lincolnshire it was used for the string courses and dressings. It was also used for many Lincolnshire churches whilst in Nottinghamshire this stone was used for the church and castle at Newark and for the carving in the chancel at Hawton, which is just south of Newark. In the time of Elizabeth I it was used for Wollaton Hall which is 2.5 miles west of Nottingham and now houses a natural history museum.

In the eighteenth century is was used at Harlaxton Manor which is just south-east of Grantham. Ancaster was also the principal stone used at Belvoir Castle.

The stone was introduced into Cambridge mainly in the nineteenth century although it was possibly used earlier at King's College Chapel. The chapel at St John's College was built using the freestone for ashlar and the weather bed for string courses and other dressings. It was used for dressings at Gonville and Caius College. The east range of Trinity Hall fronting onto Trinity Hall Lane was rebuilt in Ancaster (except for the stone surrounding the entrance). In addition, the south side of the Old Court at Peterhouse was refaced in Ancaster. These last two buildings show the typical appearance of the freestone – darker streaks running parallel to bed within individual blocks of stone.

More recent use of the stone from the present quarry can be seen at the following buildings:

Green College, Oxford – sills of the accommodation blocks, 1979
Ionic Temple, Rievaulx, Nr Helmsley – Ancaster hard white was used together with a French stone to rebuild the portico, 1980
National Westminster Bank, Grantham – restoration of dressings, 1981
Lincoln Guildhall – restoration of dressings, 1981
Gowt's Bridge, High Street, Lincoln – largely rebuilt in Ancaster, 1981
Kesteven Hospital, Nr Grantham – ashlar, 1982
Garrowby Hall, York – restoration of all dressings, 1982
St Peter's Church, Clayworth (west of Gainsborough) – hard white stone used for tracery, mullions and sills, 1982

Results of tests

Velocity of sound ($m.s^{-1}$)	Porosity (% volume)	Saturation coefficient	Microporosity (% saturation)	Capillarity coefficients French A ($g.cm^{-2}.min^{-1/2}$)	Belgian GC	Belgian G	Crystallisation test (% weight loss)
Weather bed							
4500	9.9	0.81	77	(1.6)	(4.0)	(18.5)	18, 21
4400	10.2	0.87	90	(2.2)	(3.9)	(17.9)	18, 21
Hard white bed*							
2500	15.3	0.95	79	(3.8)	(4.0)	(21.8)	All cubes
2600	14.6	0.95	80	(3.7)	(4.2)	(22.5)	tested failed
2600	14.1	0.93	–	(3.7)	(4.0)	(20.9)	before
2700	14.4	0.94	80	(5.9)	(3.6)	(21.7)	end of test
Freestone							
2900	19.2	0.84	61	(4.7)	(2.4)	(18.7)	23, 22
2900	19.3	0.83	60	(4.6)	(2.5)	(17.7)	25, 20

– Not tested
* 4 samples were tested

Durability
Weather bed: Class D
Hard white bed: Class F
Freestone: Class D

The durability classifications for the freestone and the weather bed are somewhat harsher than would be expected in the light of the reference buildings and general experience at BRS. A classification of C may be more appropriate for the average output of the weather bed.

BALLINASLOE STONE

Owners Stone Developments Ltd, PO Box 13 (Bray), Ballybrew, Co Wicklow, Eire. Tel: Ballybrew (00404) 862981 (via operator)

Quarry location Ballinasloe, Co Galway, Eire. Tel: Killimor (0905) 2119 (via operator)

The quarry is 0.5 mile to the south-west of Ballinasloe on the Galway Road — the N6. There are old quarries nearby and further areas to exploit.

Petrography *(Plate 4)*
Ballinasloe Stone is of early Carboniferous age. Under 3 m of overburden there are 14 beds of stone and the 13th and 14th are being worked. Work will begin shortly on the stone below the present level of the quarry floor. The depth of the beds is about 1 m but the depth of stone on bed is only 200 to 230 mm. There are four deep faces of stone in this large quarry. The stone is dense and smooth and is a dark blue-grey colour. It contains numerous small pieces of shell.

Reference buildings
The stone is used for monumental work, walling and cladding. The following are some examples:

Galway Cathedral — ashlar, plinths, cornices and corbels, 1966
Garden of Remembrance, Dublin — walls and fountains
University College, Dublin — cladding, 1972 to 1973
Russel Court Hotel, Dublin — ashlar, 1981

Results of tests

Velocity of sound ($m.s^{-1}$)	Porosity (% volume)	Saturation coefficient	Microporosity (% saturation)	Capillarity coefficients French A ($g.cm^{-2}.min^{-1/2}$)	Belgian GC	Belgian G	Crystallisation test (% weight loss)
5000	0.7	0.7	32	(0.04)	(0.02)	(−86.8)	0, 0
(5500)	0.3	1.00	30	0.02	0.6	18.6	0, 0

Durability
Class A

BATH STONE – COMBE DOWN – MOUNT PLEASANT QUARRY

Grid reference ST 772 624

Owners George V Williams & Sons Ltd, Windsor Bridge Road, Bath, BA2 3DT.
Tel: Bath (0225) 22286

Quarry location Mount Pleasant Quarry, Combe Down

The stone comes from the Mount Pleasant Quarry in the village of Combe Down itself. To reach it turn right into Shaft Road which is off Claverton Road (travelling eastwards). The entrance to the quarry is on the right by some cottages, about 0.5 mile from the main road, and immediately after the entrance to a hotel on the left. The quarry has only recently opened and an extension of the present planning permission is being sought before fully developing the quarry.

Petrography
Combe Down Stone is an oolitic limestone from the Great Oolite of middle Jurassic age. The stone is found under 3.6 m of overburden. Blocks of up to 1.2 m on bed are available.

Reference buildings
In 1981 the stone was used for the restoration of the balustrading of Green Park Station facade which is part of a supermarket complex in Bath.

The quarry was opened too late to be included in the BRS testing programme and to date there are no other reference buildings.

BATH STONE – COMBE DOWN – UPPER LAWNS QUARRY

Grid reference ST 766 624

Owner Mr J Hancock, 11 Beach Avenue, Claverton Down, Bath

Quarry location Upper Lawns Quarry, Combe Down, Bath.
Tel: Bath (0225) 833337

The quarry is off Shaft Road in Combe Down village. It is in St Winifreds Drive which is on the right before the entrance to the Mount Pleasant Quarry. It has been worked since 1850. In the past there were seven or eight mines and one open quarry. At the present rate of working there is a 20-year stock of stone.

Petrography *(Plate 4)*
Combe Down Stone is an oolitic limestone from the Great Oolite of middle Jurassic age. It is a buff, shelly limestone and often has characteristic veins running perpendicular to the bedding. There are several beds of stone of varying coarseness but it is difficult to tell the difference between them. The average depth of stone on bed is 450 mm, and 600 mm is the maximum obtainable.

Reference buildings
In the 1720s Woods, Nash and Allen strove to make Bath a fashionable spa to rival the European cities. Ralph Allen bought the Combe Down quarry. He developed methods of quarrying and transporting the stone so that it became more widely available. In the 1730s he built, for himself, the mansion Prior Park at Combe Down*. This was a magnificent mansion designed to show the stone at its best after he had failed to win a contract to build a London hospital because the stone was not well known. Combe Down and Box Ground (the latter is no longer available) represent the very best Bath Stone on which the reputation of stones from the area is based.

Nowadays the stone from the present quarry is used locally for housing at Fleshford. The stone has also been used to restore the minarets at Brighton Pavilion (1982) and for restoration of the north front of Westminster Hall, Palace of Westminster (1982).

Results of tests

Velocity of sound $(m.s^{-1})$	Porosity (% volume)	Saturation coefficient	Microporosity (% saturation)	Capillarity coefficients French A $(g.cm^{-2}.min^{-1/2})$	Belgian GC	Belgian G	Crystallisation test (% weight loss)
3500	25.4	0.54	34	(3.2)	(−3.6)	(−15.5)	13, 16
3500	25.7	0.55	22	(3.0)	(−3.6)	(−15.0)	13, 15

Durability
Class C

---

*The mansion is now a school – Prior Park College, Ralph Allen Drive, Combe Down. From its grounds one has a splendid view of Bath with its well laid out terraces and crescents.

BATH STONE – MONKS PARK

Grid reference ST 881 683

Owners The Stone Firms Ltd, 20 Manvers Street, Bath. Tel: Bath (0225) 61266

Mine location Monks Park, Corsham, Wilts.
Tel: Bleadon (0934) 812358 (Area Office)

The Monks Park mine is off the B3353 in Wiltshire. As you travel south, just before the village of Gastard, there is a turning to the right marked 'RN Depot and Monks Park Mine'. The mine is nearly 0.5 mile up this road and is beyond the RN depot. The stone is mined 24 m below ground in a vast mine which extends for 15½ hectares. There are plenty of reserves of stone. Only building stone is brought to the surface; any waste is returned to disused galleries. Stone is winched to the surface up an inclined shaft. It is worked in galleries about 6 m high which are supported by huge pillars of stone left untouched to act as pit props.

Nowadays cutting machines can easily make the long horizontal cut needed to release the stone from the roof. Before mechanisation, this job took days and involved the men in wielding longer and longer picks to reach in to the back of a block. At the entrance to the mine is a small museum of the type of tools used and photographs of the conditions in which the men had to work.

Petrography *(Plate 4)*
Monks Park Stone is an oolitic limestone from the Great Oolite of middle Jurassic age. It is a fine-grained, buff coloured stone. At the time of the visit four beds were visible in the 6.6 m face. The top two are shallow being 300 and 450 mm deep. The next two beds can be up to 750 mm deep. The depth of stone on bed varies from 250 to 600 mm. Then there is some waste with a further bed of building stone below. The stone from each bed looks the same.

Reference buildings
Monks Park Stone has been used extensively, often in conjunction with Westwood Ground Stone dressings. The following are some examples of both types of use.

In the 1960s Monks Park was used both for Woolworth and for Marks & Spencer in Bath. In the early 1970s it was used at the Cheltenham and Gloucester Building Society, Cheltenham, and for Police Headquarters at both Cheltenham and Cirencester. In 1973 Monks Park ashlar was used for the Sun Alliance House in Bristol. In the mid-1970s it was used for the Pump House extension at York Street, Bath. It was also used for the restoration of ashlar, columns and pedestals at the Stall Street Pump House. Here Westwood was used for the plinths, cornice and copings. Similarly, in the late 1970s Monks Park was used for the restoration of the ashlar and cornice whilst Westwood was used for the projecting courses at 1 Henrietta Street, Bath. A combination of both stones was used for restoration work at Leckhampton Court, Cheltenham, in the late 1970s and early 1980s.

The new Westminster Bank in Fore Street, Taunton, built in 1980–81, has Monks Park ashlar and Westwood copings. The First Church of Christian Science in Bristol has Monks Park ashlar and Westwood for the projecting courses. This church was built in 1981–82. In the restoration of 15 to 20 Beauford Square, Bath, carried out in 1980–81, Monks Park was used for the ashlar whilst Westwood provided the dressings.

(continued)

Further afield, at Kings Road, Reading, Monks Park ashlar and Westwood dressings were used in 1981 for the restoration of houses at numbers 179 and 181–183 and for the new office block at number 185. Currently the two stones are being used for the National Farmers' Union and Avon Mutual Insurance Co building in Stratford-on-Avon. This building is off the B4086 south of the river.

Results of tests   Bath Stone – Monks Park Stone

Velocity of sound (m.s^{-1})	Porosity (% volume)	Saturation coefficient	Microporosity (% saturation)	Capillarity coefficients French A (g.cm^{-2}.min$^{-1/2}$)	Belgian GC	Belgian G	Crystallisation test (% weight loss)
Top bed							
(3200)	21.1	0.95	82	5.8	4.5	22.8	52, 68
2900	21.9	0.95	85	(5.9)	(4.6)	(24.4)	–
(2700)	24.7	0.94	–	8.6	4.0	24.2	49, 52
(2600)	26.7	0.96	–	7.0	4.2	23.4	–
Bottom bed							
3100	19.5	0.96	–	(4.8)	(4.4)	(24.4)	–
(3200)	19.8	0.94	85	5.1	3.9	22.3	68, 55
3000	22.8	0.99	77.5	(10.0)	(4.3)	(26.9)	69, 72
(3100)	21.4	0.96	–	–	–	–	–
Market bed							
(2900)	22.3	0.94	77	6.1	4.3	24.1	1 cube failed, 59
2800	24.5	0.98	–	(12.9)	(3.8)	(26.9)	–
2800	22.9	0.93	83	(6.2)	(4.6)	(24.6)	68, 75
(2800)	22.8	0.94	–	6.3	4.6	24.8	–

– Not tested

Durability
Top bed: Class E
Bottom bed: Class E
Market bed: Class E to F

BATH STONE – STOKE GROUND BASE BED

Grid reference ST 776 597

Owners The Bath Stone Co Ltd, Canal Cottage, Avoncliffe, Bradford-on-Avon, Wiltshire, BA15 2HD. Tel: Bradford-on-Avon (022-16) 5260

Mine location Hayes Wood Mine, Limpley Stoke, Wiltshire

The mine is in Midford Lane, Limpley Stoke, just off the A36 south of Bath. The mine was last worked in 1940 when a stone known as Stoke Ground was available. The mine was used as a munitions store during the war. It was reopened at the beginning of 1982.

Petrography *(Plate 4)*
Stoke Ground Base Bed is an oolitic limestone from the Great Oolite of middle Jurassic age. The stone is mined approximately 15 m below ground but access is relatively easy. There are two beds of stone but only one, the bottom bed, is currently available. This is a creamy-buff coloured stone with a fairly coarse structure and many small pieces of shell. The depth of stone on bed is about 1 m and the total depth of the bed is not much more than this.

The bottom bed was not worked previously because it was too hard for the tools used in the 1930s. The present company intends to work all of this untouched bed until they reach the point in the face at which the mine was abandoned. They will then also work the top bed – the Stoke Ground. It will be approximately 2 years before the latter bed is available.

The mine was reopened too late to be included in the full BRS test programme. However, the company can supply independent test results on request.

Reference buildings
This stone was used for the carved caps to the columns at Drapers Hall, Coventry. It was used at Windsor Castle for chimney stacks and caps by the private apartments. At Lambeth Palace, London, it was used for chimney stacks, and at Hampton Court Palace for copings. All of these jobs were completed in 1982.

Results of tests

Porosity (% volume)	Saturation coefficient	Crystallisation test (% weight loss)
19.3	0.73	30, 31
		16, 13

Durability
Class D

BATH STONE – WESTWOOD GROUND

Grid reference ST 803 591 (Bradford-on-Avon)

Owners The Stone Firms, 20 Manvers Street, Bath. Tel: Bath (0225) 61266

Mine location Westwood Quarry, Westwood, Nr Bradford-on-Avon.
    Tel: Bleadon (0934) 812358 (Area Office)

The mine is in the village of Westwood. It is on the hill that looks down into the valley of the River Avon and onto Bradford-on-Avon beyond. The entrance to the mine is on the side of the hill and you walk in to the face which is below ground. The mine was reopened in 1975 and there are plenty of reserves of stone.

Petrography *(Plate 4)*
Westwood Ground Stone is an oolitic limestone from the Great Oolite of middle Jurassic age. It is a coarse-grained, buff coloured stone. There are two beds of stone in a 3 m deep face. The top bed is between 600 and 750 mm deep whilst the lower bed, which is harder, is up to 1.2 m deep. The stone below the lower bed has clay holes and is cut off as waste. Blocks of stone can be supplied up to 1.1 m on bed.

Reference buildings
Westwood Stone is often used to provide the dressings where Monks Park Stone has been used as ashlar. The reader should also refer to the Monks Park reference buildings where several examples of the use of Westwood are quoted. The following are some additional examples.

Westwood was used for the general restoration work carried out, in 1978–79, at Grangetown and Lansdowne Primary School, Cardiff. At about the same time it was used for the window panels and copings at Clifton Down Station, Bristol (Monks Park provided the ashlar). It was used for the restoration of jambs, plinths and copings at All Saints Church, Fulham in 1979–80. The new Halifax Building Society at Witney, built in 1980–81, has Westwood sills, copings, heads and fascia. At the same time it was used for new dormer windows at Fetcham Park House, Fetcham, Leatherhead.

Westwood was used for the replacement of the top 3.6 m of the church spire at Todenham, Gloucester. It provided all the dressings for a block of 22 flats at Dollar Street, Cirencester. It was used for the spire tower replacement at St Peter's Church, Bournemouth, in 1981–82.

Results of tests

Velocity of sound (m.s^{-1})	Porosity (% volume)	Saturation coefficient	Microporosity (% saturation)	Capillarity coefficients French A (g.cm^{-2}.min$^{-1/2}$)	Belgian GC	Belgian G	Crystallisation test (% weight loss)
Top bed							
3000	20.7	0.93	74	(4.7)	(4.0)	(22.8)	60, 61
(3000)	20.7	0.90	76	4.3	3.2	17.8	–
(3100)	19.2	0.91	–	4.1	3.6	21.5	–
3200	19.1	0.85	–	(3.7)	(2.6)	(17.5)	22, 20
Bottom bed							
3000	19.2	0.97	–	(4.3)	(4.6)	(27.1)	–
(3100)	19.7	0.93	59	4.5	4.0	23.5	54, 52
(3000)	21.4	0.94	–	4.2	3.9	21.3	–
3100	21.1	0.96	57	(4.8)	(3.8)	(21.2)	53, 58

– Not tested

Durability
Top bed: Class E
Bottom bed: Class E

The durability classification is harsher than would be expected in the light of test results obtained by BRS over the years. Stone of Class D is usual.

CLIPSHAM STONE

Clipsham was first used by the Romans for their villas. It was used locally for churches and was also sent to Windsor Castle in the fourteenth century. But, it is the hardest of the Lincolnshire Limestones and was not extensively used until the introduction of frame saws in the seventeenth century. Nevertheless, it had a good reputation and despite the difficulties in cutting it the stone was used sporadically, eg at King's College Cambridge in the sixteenth century along with Weldon and a magnesian limestone.

However it was the city of Oxford, and not the geographically nearer Cambridge, which made most use of Clipsham for building. Sir Thomas Jackson introduced the stone to Oxford in the 1870s when it was used for the Examination Schools. Jackson went on to use Clipsham for the dressings of the New Quadrangle at Brasenose College. At Lincoln College the Grove buildings and a wing south-east of the Rector's Lodge were both given Clipsham dressings.

The Clipsham was used in combination with the local Bladon stone which provided a walling stone. Some Oxford buildings with Clipsham ashlar include the Arlosh Hall of Manchester College (1915), the chapel at Somerville (1934), the Playhouse in Beaumont Street (1938) and the south corner buildings at Carfax (1931 – 32).

Many of Oxford's buildings were built or faced with local Headington stone. Much of this proved to be a poor-quality stone and already in the nineteenth and early twentieth centuries there was a need for restoration work. Jackson used Clipsham in 1873 to repair some of the pinnacles of the Bodleian (Old Schools) quadrangle. Between 1902 and 1906 Hawksmoor's twin towers at All Souls were rebuilt, as was the top storey of The Tom Tower and the west front of Christ Church in 1909 – 12. In each case Clipsham was used.

More recently (1962) Clipsham was used for the capitals and moulded band courses at Christ Church Library. The stone was also used extensively for the repairs to the Sheldonian Theatre carried out between 1958 and 1961 and to the Old Ashmolean Building at about the same time. There are many more Oxford buildings where Clipsham has been used for restoration work, and today this is probably the stone most in demand for repairs and restorations.

The use of Clipsham is not restricted to Oxford. Since the 1920s it has been used extensively as the main replacement stone for the Palace of Westminster and has also been used at Buckingham Palace and Hampton Court.

It has been used for repairs and restoration at cathedrals as far apart as Canterbury and York, Ripon and Salisbury, as well as Peterborough, Ely and Norwich. At Norwich it is replacing decayed French stone.

More recent Clipsham buildings are Nuffield College and the Inorganic Chemistry Laboratory at Oxford (late 1950s). The Berkeley Hotel, Knightsbridge, has a Clipsham facade built in 1973. Winchester Law Courts, built in 1974, have Clipsham dressings and quoins on the north elevation of the Court House.

At the time of the survey only one quarry – Medwells – was actively extracting building stone. In addition, stone from the once famous quarries of Big Pits and Longdales was being sorted and offered for sale. Stone from all three quarries was obtained for testing.

CLIPSHAM STONE – BIG PITS AND LONGDALES QUARRIES

Grid references Big Pits SK 967 146, Longdales SK 976 153

Owners Clipsham Quarry Co, Clipsham Hall, Clipsham, Oakham, Leics. Tel: Castle Bytham (078-081) 204

Agents Murrays, Market Street, Oakham, Leics. Tel: Oakham (0572) 55555

The Clipsham Quarry Co is offering stone from two of the four Clipsham quarries that it owns. These are the Big Pits and Longdales quarries. Big Pits was opened in the eleventh or twelfth century and was used sporadically. The present company opened it in 1923 and closed it in 1952. Longdales was abandoned in the same year having been opened in 1938.

Big Pits can be reached by turning right just before the village of Clipsham when travelling from the A1. Follow the road past the farm to the quarry entrance on the left side of the road.

Longdales is to the right of the road between Clipsham and Careby. The entrance is marked 'Soil Fertility Quarry' and Longdales is further along this track.

Petrography *(Plate 4)*
Clipsham Stone is an oolitic limestone from the Lincolnshire Limestone formation of middle Jurassic age. It is a creamy-brown coloured stone with many pieces of shell. Some blocks can contain substantial amounts of blue stone. In the 1930s and early 1950s large quantities of stone were quarried for use at York Minster and the Palace of Westminster. Stone of exceptional quality was required. Blocks which did not conform to the demanding specifications were put back and this is the stone that is currently for sale. Blocks were put back randomly and would have come from different parts of the quarry. Blocks from Longdales contain a large proportion of blue stone.

The company owns a further two quarries from which 'put back' blocks may also be offered in the future.

Reference buildings
To date, substantial quantities of stone have been sent for a new shopping development in Peterborough.

(continued)

Results of tests Clipsham Stone

Velocity of sound (m.s^{-1})	Porosity (% volume)	Saturation coefficient	Microporosity (% saturation)	Capillarity coefficients French A (g.cm^{-2}.min$^{-1/2}$)	Belgian GC	Belgian G	Crystallisation test (% weight loss)
Big Pits Quarry – Creamy-brown stone							
3700	14.6	0.92	79	(4.5)	(3.8)	(24.6)	21, 17
3500	14.9	0.92	78	(5.0)	(3.3)	(22.6)	17, 20
3400	14.8	0.90	68	(5.3)	(2.8)	(19.6)	15, 19
(3500)	14.1	0.88	57	3.6	3.0	19.0	25, 21
(2100)	19.0	0.87	50	10.0	1.8	16.7	40, 35
2100	18.3	0.89	43	(7.7)	(2.2)	(19.0)	33, 33
Longdales Quarry – Brown stone							
2800	14.9	0.77	75	(3.6)	(1.4)	(11.5)	20, 13
3200	16.3	0.88	76	(5.5)	(3.0)	(22.5)	23, 26
Longdales Quarry – Blue stone							
4000	12.5	0.78	76	(1.0)	(2.3)	(18.3)	0, 0
3800	11.5	0.87	82	–	–	–	0, 0

– Not tested

Durability
Big Pits Quarry – Creamy-brown stone: Class D
Longdales Quarry – Brown stone: Class D
              Blue stone: Class A

Although all the samples of creamy-brown stone and brown stone tested for the survey were only Class D, the durability of the stone on which the excellent reputation of Clipsham is based would certainly be higher.

CLIPSHAM STONE – MEDWELLS QUARRY

Grid reference SK 987 158

Owners and quarry location P G Medwell and Son, The Quarries, Clipsham, Oakham, Leics. Tel: Castle Bytham (078-081) 224

This quarry, opened in 1903, is to the left of the road between Clipsham and Careby about 1 mile from Clipsham to the north of Stamford. It is the second of the two entrances and is opposite some houses and farm buildings. It backs onto the medieval Holywell Quarry which is no longer worked. There are plenty of reserves of stone.

Petrography *(Plate 4)*
Clipsham Stone is an oolitic limestone from the Lincolnshire Limestone formation of middle Jurassic age. It is a creamy-brown coloured stone with many pieces of shell. Some blocks can contain substantial amounts of blue stone. At the time of the visit there was a separate face of predominantly blue stone with blocks of about 1 m on bed. Whilst blocks of creamy-brown stone of up to 1.8 m are obtainable, 900 to 1200 mm is the normal depth on bed. The stone is found under a great deal of overburden – up to 15 m in some parts of the quarry.

Reference buildings
This quarry was the only source of Clipsham Stone from the early 1950s until the present day. Thus it is the Medwells Quarry that has supplied the new stone for the Inorganic Chemistry Laboratory and Nuffield College at Oxford and the Peers' Entrance at the Houses of Parliament, and, of course, all the Clipsham for recent restorations at Oxford, Cambridge and Norwich. This stone was also used in 1979 for string courses, architraves, cornice, coping and cap of the gateway to Green College Oxford, and in the courtyard of this college Clipsham was used for the arch stones, string courses and ashlar on the east elevation.

Results of tests

Velocity of sound $(m.s^{-1})$	Porosity (% volume)	Saturation coefficient	Microporosity (% saturation)	Capillarity coefficients French A $(g.cm^{-2}.min^{-1/2})$	Belgian GC	Belgian G	Crystallisation test (% weight loss)
Creamy-brown stone							
3200	25.0	0.85	75	(2.1)	(2.6)	(17.6)	14, 16
(2900)	12.5	0.87	78	3.0	3.2	19.9	19, 27
Blue stone							
4300	9.6	0.94	82	(1.0)	(4.4)	(24.8)	1, 1

Durability
Creamy-brown stone: Class D
Blue stone: Class A

Although the classification of the creamy-brown stone tested for the survey is only Class D, the stone on which the excellent reputation of Clipsham is based would certainly be higher.

DAGLINGWORTH STONE

Grid reference SP 001 059

Owners Kingston Minerals Ltd, Midland Region, Whelford Road, Fairford, Glos. Tel: Cirencester (0285) 712471

Quarry location Daglingworth, NW of Cirencester, Glos. Tel: Cirencester (0285) 5961

The quarry is 2 miles north-west of Cirencester just off the A417. It is an extremely large quarry and there are vast reserves of stone. The quarry was opened after the Second World War but building stone has only been extracted since about 1955. Before that all of the output was crushed for aggregates or for making reconstituted stone. Indeed, most of the output is still crushed although reconstituted stone is no longer produced.

Petrography *(Plate 4)*
Daglingworth Stone is an oolitic limestone from the Great Oolite of middle Jurassic age. It is a light buff colour with very small speckles of a darker colour. At the time of our visit building stone was only being extracted from one part of the quarry. There is a small amount of overburden followed by 1.8 m of rag or brash stone. Then there are 2.4 to 3.0 m of randomly bedded stone. Within the latter layer is approximately 600 mm of building stone. This building stone is found as 50 to 75 mm and 100 to 125 mm blocks on bed. The stone is lifted off by excavator and sorted – there is no need to use plug and feathers to extract it. Only small blocks can be provided as there is no machinery for cutting larger blocks.

There are further beds of stone below the building stone but this too is crushed.

Reference buildings
Daglingworth Stone is generally used as a walling stone. In the 1960s it was used in an external feature wall at Princess Margaret Hospital, Swindon. In the same decade it was used for feature panels in a housing estate at Chippenham Road, Calne, Wilts. The stone was used for two bridges on the M50, which was opened in 1960. It was used for the paved area under the bridge where the motorway crosses the B4211. It was also used for the abutments on the Ripple Road and Railway Bridge which is the first bridge as you go west from Junction 1. In 1979 the courtyard of Green College Oxford was paved in Daglingworth with York Stone used for the surrounds.

There is a Daglingworth Stone retaining wall on the verge of Gloucester Road, Cirencester, built in 1980.

Results of tests

Velocity of sound (m.s^{-1})	Porosity (% volume)	Saturation coefficient	Microporosity (% saturation)	French A (g.cm^{-2}.min$^{-1/2}$)	Belgian GC	Belgian G	Crystallisation test (% weight loss)
(4800)	4.7	0.89	–	0.6	4.3	18.4	–
4500	4.3	0.88	72	(0.5)	(4.8)	(5.3)	3, 0
(4800)	5.5	0.86	70	0.9	3.9	17.9	11, 12
5300	3.5	0.83	–	(0.3)	(2.4)	(19.0)	–

– Not tested

Durability
Class C

DOULTING STONE

Grid reference ST 648 436

Owners The Stone Firms Ltd, 20 Manvers Street, Bath. Tel: Bath (0225) 61266

Quarry location Doulting, Nr Shepton Mallet, Somerset

The present Doulting quarry is located at the village of Doulting on the A361 near Shepton Mallet. There are old famous quarries nearby like Brambleditch and Chelynch and the stone has been worked since Roman times. Although there is a large potential reserve of stone the quarry was not being worked on a continuous basis at the time of the visit.

Petrography *(Plate 5)*

Doulting Stone is from the Inferior Oolite of Jurassic age. Unlike the stone extracted centuries ago, the currently available stone is rarely oolitic. It is composed of fragments of older Carboniferous (or perhaps Liassic) limestones, which were eroded and later redeposited, resulting in a crystalline and coarsely granular appearance. This distinguishes Doulting Stone from other Jurassic limestones.

The building stones lie under a small amount of overburden. There are four beds as follows:

Top bed – very crystalline to the extent that the stone almost seems to sparkle. The stone is cream to mottled brown in colour and the depth of the bed is 600 mm.

Rag bed – also very crystalline but the stone is more coarse than the top bed and is creamy brown in colour. The depth of the bed is 450 to 600 mm.

Grey bed – less crystalline than the above two beds with a finer grain and more cream in colour than brown but certainly not grey. This bed is regarded as the bottom of the rag bed and is 300 to 450 mm deep.

Bottom bed – another crystalline stone with large coarse grains, mostly brown in colour with little cream. The depth of this bed is 750 mm.

The depth of stone on bed varies. As the quarry face is worked back, two additional beds are sometimes uncovered. One occurs on top of the four recognised beds mentioned above and one below them.

Reference buildings

The most famous examples of buildings where Doulting Stone can be seen are Glastonbury Abbey founded in the twelfth century and Wells Cathedral largely erected in the thirteenth century. Most of the current output from the quarry is sent to Wells for restoration work.

The stone has been used at the All Saints Russian Orthodox Church, Ennismore Gardens, London. This church was originally constructed in brick but in 1892 the west front was faced in Doulting. In the 1930s the stone was used for the reception area of the county offices, Taunton. In the 1950s Doulting was used for dressings and the nave at Guildford Cathedral. In the 1960s it was used for the facings to the two bridges at Warndon, which is at Junction 6 on the M5 near Worcester. In 1973 it was used to face the river-side walls at the Pulteney Weir scheme, Bath, and for eight columns supporting the vaulting and for the ribs of the west crypt of the Guildhall, London. The infill panels are of Portland. In 1976 it was used, with Purbeck Limestone, at the new town centre in Shepton Mallet. The ashlar, copings and dressings are largely in Doulting. In 1981 it was used at Lancing College Chapel, Sussex, for one new bay with a vaulted ceiling and rose window.

(continued)

Doulting frequently becomes grey on weathering and its final appearance may be very different from its original creamy-brown colour. It sometimes shows a characteristic weathering pattern, within individual blocks, in which weak bands weather out faster than the surrounding stone.

Results of tests   Doulting Stone

Velocity of sound (m.s⁻¹)	Porosity (% volume)	Saturation coefficient	Microporosity (% saturation)	Capillarity coefficients French A (g.cm⁻².min⁻¹ᐟ²)	Belgian GC	Belgian G	Crystallisation test (% weight loss)
Top bed							
2600	11.9	0.63	22	(1.9)	(−2.6)	(−7.7)	6, 5
(1900)	22.5	0.76	−*	12.0	−1.2	7.5	38, 34
Rag bed							
(2200)	14.6	0.60	30	1.8	−2.5	−6.4	4, 3
2300	16.9	0.63	−*	(2.8)	(−2.7)	(−7.6)	4, 6
Grey bed							
(2900)	12.2	0.72	52	2.4	−0.8	0.9	14, 12
2900	12.2	0.69	48	(2.4)	(−0.9)	(−0.3)	8, 10
Bottom bed							
(2200)	12.8	0.69	17	2.2	−1.5	1.6	7, 4
2500	12.7	0.74	19	(4.3)	(−1.3)	(4.8)	6, 8

*Stone too soft to give a satisfactory sample for microporosity test

Durability
Top bed: Class D
Rag bed: Class B
Grey bed: Class C
Bottom bed: Class C

FENACRE STONE

Grid reference ST 060 172

Owners and quarry location King's Asphalt Ltd, Fenacre Quarry, Burlescombe, Nr Tiverton, Devon. Tel: Greenham (0823) 672206

Petrography *(Plate 5)*

Fenacre Stone is of early Carboniferous age. It is a black, shiny stone with white and red veins running through it. There is one large face of stone 45 m deep below 600–900 mm of overburden. Within the face the depth of the beds varies from 50 mm to about 1 m.

Reference buildings

The stone is used locally for housing, sea walls and bridge works but mainly to provide roadstone. It is extracted from the same hill as Westleigh Stone but from a different side. Westleigh Stone was quarried many years ago.

Results of tests

Velocity of sound (m.s^{-1})	Porosity (% volume)	Saturation coefficient	Microporosity (% saturation)	Capillarity coefficients French A (g.cm^{-2}.min$^{-1/2}$)	Belgian GC	Belgian G	Crystallisation test (% weight loss)
5200	0.8	0.76	32	(0.04)	(−4.6)	(−4.04)	0, 0
5700	0.4	0.40	17	(0.002)	(−5.2)	(−58.5)	0, 0

Durability

Class A

GUITING STONE – COSCOMBE QUARRY

Grid reference SP 079 302

Owners Kingston Minerals Ltd, Whelford Road, Fairford, Glos.
Tel: Cirencester (0285) 712471

Quarry location Coscombe Quarry, Nr Ford, Glos.

This extremely large quarry is on the B4077 just north of Ford which is north-west of Stow-on-the-Wold. There are plenty of reserves of stone.

Petrography *(Plate 5)*
Guiting Stone is an oolitic limestone from the Inferior Oolite of middle Jurassic age. Samples of seven different stones were obtained as follows:

White walling stone — light-cream coloured stone of which blocks 150 to 230 mm on bed can be obtained. The total depth of this bed is 3 m.

Top of ledge } both are a darker cream than the walling stone but are currently
Bottom of ledge } crushed and not available as building stone.

Top of masonry — dark cream stone which is 300 to 450 mm on bed.

Middle of masonry — yellow-orangey stone which is 300 to 450 mm on bed.

Bottom of masonry — yellow-orangey stone which is up to 600 mm on bed.

Floor — dark cream stone which is up to 600 mm on bed.

Stone from the masonry and floor areas forms a face 7.2 m in depth, there being several beds of each type of stone within this face. Stone from these beds is collectively described, by the quarry, as dark cream. However, stone from the middle and bottom of masonry beds is a very distinctive orange. All the beds provide a coarse stone containing many pieces of shell.

Reference buildings
Guiting is one of the few remaining quarries working Cotswold limestone. In previous centuries nearly every Cotswold community had its own quarry to provide the stone for the numerous beautiful stone villages whose character is carefully preserved and protected today. Many of the stones were only used locally and are not to be found outside the district. However, there are some examples of Guiting Stone buildings further afield.

The early Tudor Gatehouse close to the Abbey in Tewkesbury was built in Guiting. The addition to the west range at Balliol (facing St Giles Church in Oxford) has a Guiting facade and was built in 1906–7. The Royal Oxford Hotel, near the station, was built in 1938 with Guiting ashlar and dressings. A current contract in Oxford involves the restoration of the Bursar's Lodge at Magdalen College.

In 1970 a new chapel within the grounds of Prinknash Abbey in Gloucestershire was built with Guiting. In the centre of Abingdon near the ruin of the Abbey by the river the stone was used for the residential flats known as Cosenor's House (1980). In the main street of Clipsham in Lincolnshire four houses were built in rough dressed Guiting in 75, 150 and 230 mm deep blocks (1980). Police houses in Sturminster Newton, Dorchester, are also of this stone (1981). In 1981 a small estate of houses at Hunton-in-the-Hedges near Brackley was built in Guiting. All of the above reference buildings were built in the yellow-orangey Guiting.

The following are in the white walling stone:
The National Westminster Bank, Witney — 1978
Stratton Audley Manor, Bicester — housing block within the grounds, 1979
Tesco Superstores, Cirencester — reconstructed front to the store, 1980
Halifax Building Society, Witney — 1981

Results of tests

Velocity of sound (m.s^{-1})	Porosity (% volume)	Saturation coefficient	Microporosity (% saturation)	Capillarity coefficients French A (g.cm^{-2}.min$^{-1/2}$)	Belgian GC	Belgian G	Crystallisation test (% weight loss)
White walling							
(3600)	15.6	0.90	78	3.6	3.9	20.1	–
3400	15.7	0.96	75	(4.4)	(4.7)	(27.3)	73, 73
3600	15.1	0.89	–	(3.1)	(4.1)	(22.3)	–
(3500)	16.3	0.96	–	4.7	4.2	22.7	–
Top of ledge							
(3000)	25.4	0.80	–	6.9	1.8	19.9	–
3000	24.5	0.83	76	(6.3)	(2.8)	(21.4)	52, 45
(3000)	23.7	0.84	74	4.5	3.3	19.3	43, 47
3000	25.7	0.85	–	(8.4)	(2.7)	(17.6)	–
Bottom of ledge							
(3200)	19.5	0.91	–	3.3	3.5	19.2	67, 48
3200	21.5	0.88	–	(3.5)	(3.1)	(17.9)	–
4000	–	–	83	(7.3)	(0.8)	(23.7)	–
(3400)	18.9	0.89	84	3.9	4.0	21.1	34, 46
Top of masonry							
3000	26.8	0.77	–	(8.7)	(2.3)	(17.7)	–
(3300)	18.5	0.89	76	4.7	3.9	21.3	20, 27
(2500)	21.0	0.91	–	6.6	2.2	24.4	–
3200	18.6	0.94	71	(5.2)	(4.1)	(25.5)	32, 38
Middle of masonry							
(2500)	25.6	0.84	66	11.8	1.4	16.0	24, 16
2900	24.7	0.84	66	–	–	–	–
(2500)	24.9	0.81	–	4.6	2.0	18.7	–
2900	24.4	0.82	–	(7.9)	(1.7)	(16.5)	26, 24
Bottom of masonry							
(3400)	20.1	0.78	54	5.2	0.5	9.1	11, 11
2900	26.1	0.69	–	(5.0)	(−0.6)	(−0.3)	–
(2600)	26.1	0.87	51	12.4	1.7	18.7	27, 35
Floor							
3000	19.9	0.87	–	(6.4)	(2.6)	(21.3)	–
(3100)	18.7	0.85	51	4.7	2.5	22.5	25, 23
3000	21.3	0.79	54	(6.4)	(1.1)	(13.6)	16, 19
(2900)	19.3	0.85	–	6.0	2.4	21.0	–

– Not tested

Durability

Walling: Class E
Top of ledge: Class E
Bottom of ledge: Class E
Top of masonry: Class D
Middle of masonry: Class D
Bottom of masonry: Class D
Floor: Class D

The test results for the three masonry beds indicate Class D stone; previous BRS results show that Class C stone may sometimes be obtained.

GUITING STONE – COTSWOLD HILL QUARRY

Grid reference SP 086 294

Owners and quarry location Palmer & Beetson Ltd, Cotswold Hill Quarry, Ford, Nr Temple Guiting, Cheltenham, Glos.
Tel: Stanton (038-673) 493

This quarry is at Ford which is north-west of Stow-on-the-Wold on the B4077. It is a new quarry but there are old workings nearby. It was opened in 1981 which was too late to be included in the full testing programme but some results are given.

Petrography *(Plate 5)*
Guiting Stone is an oolitic limestone from the Inferior Oolite of middle Jurassic age. The overall depth of the face is 24 m. There is a large amount of overburden within which is a creamy coloured bed of walling stone. There are then five beds of building stone of which only the middle three are supplied as follows:

Bed 2 – 1.2 m to 1.5 m deep of a cream-yellow stone.

Bed 3 – 1.8 to 2.1 m deep of a cream-yellow stone.

Bed 4 – about 1.2 m deep of a dark yellow-orange stone.

The average depth on bed is about 600 mm although as the face is being worked back the depth of the beds is increasing.

Reference buildings
It is too early for this stone to have been used in many contracts. However, it was used for the ashlar at Honington Bridge near Shipston on Stour, Warwickshire, in 1982.

Results of tests

Porosity (% volume)	Saturation coefficient	Crystallisation test (% weight loss)
Bed 2		
16.2	0.84	15, 15, 11
18.2	0.89	16, 17, 13
Bed 3		
25.6	0.66	17, 17, 19
25.7	0.66	17, 14, 17
Bed 4		
19.5	0.75	28, 35, 26
18.5	0.78	26, 35, 37

Durability
Bed 2: Class C to D
Bed 3: Class D
Bed 4: Class D

HAM HILL STONE

Grid reference ST 478 172 (for the north of Hamdon Hill)

Owner Mr R England, Little Follys, Ash, Martock, Somerset.
Tel: Martock (0935) 822665

Quarry location Stoke sub Hamdon, Nr Martock

The present quarry is on the north side of Hamdon Hill at Stoke sub Hamdon. The stone has been worked since Roman times and there are old quarries nearby. The stone is quarried to order and is not worked on a continuous basis.

Petrography *(Plate 5)*
Ham Hill Stone is from the Upper Lias division of Jurassic age. It is an attractive stone, yellow-brown in colour with a mottled appearance due to the presence of iron. Many pieces of shell are evident. There is a 4.5 m deep face under 9 m of overburden. There are false beds resulting in some wastage but the depth on bed is 600 mm and large blocks are available.

The stone is known as grey Ham Hill. Yellow Ham Hill is only available from the south of Hamdon Hill. Currently, there is no planning permission to work the south side although there is a possibility that a quarry could be opened this year (1983). Readers should consult the *Natural stone directory* for up-to-date information.

Reference buildings
In the past both grey and yellow Ham Hill Stone were used together as in nearby Montacute House built in the seventeenth century. The yellow stone was easier to work and tended to be used for mullions and any carved work. The grey is said to be more durable, which visual evidence tends to support.

Earlier, in the 1440s, Ham Hill dressings were used at the Hospital of St John, Sherborne. In the village of Somerton, which is about 7 miles north of the quarry, there are Ham Hill stone arcades inside the church, whilst the windows of the Town Hall and Hext Almshouses are also of Ham Hill. These buildings are from the seventeenth and eighteenth centuries. In the mid-seventeenth century the stone was used at Forde Abbey, Dorset, which is 3 miles south-east of Chard.

However the stone was mostly used in the nineteenth century. In Cambridge it was used to remodel the side of the Master's Lodge at Trinity College which faced onto the Library Court. It was also used for the dressings of Westcott House in Jesus Lane. And of course, the stone was used extensively in south Somerset and in nearby parts of Dorset. It was used at West Coker Manor, which is 2 miles south-west of Yeovil, for the roof and walls. Ham Hill was often used together with local blue Lias Stone. Both stones can be seen at the churches of Isle Abbotts, Huish Episcopi and North Petherton.

The stone has been used further afield at Hamstone House, Weybridge, Surrey, and for parts of St Anne's Cathedral, Belfast. In 1910 Ham Hill was used together with bands of flint for the refacing of the south-west wing of Chantmarle near Cattistoke in Dorset. In 1981 it was used for a new shop front at 4–5 Fore Street, Taunton. It has also been used for restoration work at St John's Church, Yeovil, and Chideock Church near Bridport.

(continued)

Results of tests   Ham Hill Stone

Velocity of sound (m.s^{-1})	Porosity (% volume)	Saturation coefficient	Microporosity (% saturation)	Capillarity coefficients French A (g.cm^{-2}.min$^{-1/2}$)	Belgian GC	Belgian G	Crystallisation test (% weight loss)
3500	17.5	0.74	38	(2.8)	(0.7)	(14.0)	1, 0
4000	13.5	0.65	39	(1.8)	(−1.6)	(0.6)	1, 1

Durability
Class A to B

HAPPYLANDS STONE

Grid reference SP 119 369

Owners OTS Holdings Ltd, Springhill, Nr Moreton-in-Marsh, Glos

Quarry location Broadway, Glos. Tel: Broadway (0386) 853409

The quarry is just off the A44, 1 mile south of Broadway. At the time of the visit the quarry had been open for 10 years but is not currently being worked. There are good reserves of stone.

Petrography *(Plate 5)*
The stone from Happylands is a yellow-buff coloured oolitic limestone from the Inferior Oolite of middle Jurassic age. There is a total depth of 45 m of stone although only the top 3 to 4.5 m were worked. Under about 1 m of overburden there is 900 mm of good stone which has been followed and worked out over the entire quarry. There is then further waste until a 3 m face is reached. It is from here that the stone was extracted. The beds of stone are interwoven making extraction difficult. The maximum depth of stone on bed is 900 mm.

Reference buildings
The stone is used locally for housing at Broad Campden.

Results of tests

Velocity of sound (m.s^{-1})	Porosity (% volume)	Saturation coefficient	Microporosity (% saturation)	Capillarity coefficients French A (g.cm^{-2}.min$^{-1/2}$)	Belgian GC	Belgian G	Crystallisation test (% weight loss)
3200	20.3	0.89	83	(8.4)	(2.5)	(15.9)	Both cubes failed after 10 cycles
(3100)	22.9	0.86	70	7.3	2.0	12.3	1 cube failed after 8 cycles and the other after 10 cycles

Durability
Class F

HOPTON WOOD STONE

Grid reference SK 282 545

Owners Tarmac Roadstone (Eastern) Ltd, John Hadfield House, Dale Road, Matlock, Derbyshire

Quarry location Middle Peak Quarry, Nr Wirksworth, Derbyshire

This quarry is just to the north of Wirksworth and lies to the west of the B5023. Most of the output from the quarry is blasted for crushing, but towards the end of 1981 work began to make available some stone for building use. The company offering the building stone is a wholly owned subsidiary of Tarmac — Frank England and Co, Bellmoor, Retford, Notts; Telephone: Retford (0777) 703891. There are old quarries nearby and Hopton Wood was used extensively in the past.

Petrography *(Plate 5)*
Hopton Wood Stone is of early Carboniferous age. The stone is usually cream or grey and many attractive fossils are present. Work is concentrated at one particular point and is progressing back away from the face where the stone used to be fractured by blasting. The stone is currently being extracted by plug and feathers and not by blasting. The depth on bed is 1.5 to 1.8 m and this may increase as the face is worked back.

Reference buildings
The quarry has not been working long enough to enable the inclusion of any recent contracts. However much of the current output is being used for monumental work and cladding.

In the past Hopton Wood Stone was often used as a decorative material and for flooring. It was used, together with an Italian marble, for the floor of the great hall at Kedleston Hall which is north-west of Derby. It can be seen at the Guildhall (1789), the Victoria and Albert Museum (1857), and the Geological Museum (1935) — all in London. In 1820 it was used for enlarging Chatsworth. It can also be seen at the Town Hall, Manchester.

More recently, in 1980, some stone was especially made available for flooring at Birmingham Cathedral.

Results of tests
Hopton Wood Stone was not available in time to be included in the test programme. However, the average values for stone tested over the years at BRS are as follows:

Porosity 7.6%
Saturation coefficient 0.61

HORNTON STONE

Grid reference SP 375 471

Owners Hornton Quarries Ltd, Edgehill, Banbury, Oxfordshire.
Tel: Edgehill (029-587) 238

Quarry location Ratley, Nr Warmington, Warwickshire

The quarry is between the A422 and the B4086 at the village of Ratley. The stone has been quarried since the eleventh century, initially in the village of Hornton itself, but quarrying operations have now moved to Ratley.

Petrography *(Plate 5)*
Hornton Stone is from the Middle Lias division of Jurassic age. It is greenish-blue or brown or can be a combination of these colours. The average depth of stone on bed is 450 mm; blocks up to 750 mm can be obtained but these are exceptional. Stone is merely lifted out by excavator in blocks which are generally 1.2 to 1.8 m in length.

Reference buildings
Hornton Stone has been quarried for centuries; it was used at Ratley Church which is mentioned in the Domesday Book and for the fourteenth-century manor house Broughton Castle which is south-west of Banbury. More recently it was used for the staircase of Cambridge University library. The front of the Welsh House in Market Square, Northampton, which was restored in 1972, is in Hornton. There is a Hornton stone archway at Wroxton College in the nearby village of Wroxton erected in 1974. It was used for stone panels at Warwickshire County Council offices, as cladding for the library in Worcester and for rustic walling at the Stratford-on-Avon Hilton Hotel. It has also been used for the reredos at Ampleforth Abbey Church, Yorkshire. It is used locally for housing.

Results of tests

Velocity of sound ($m.s^{-1}$)	Porosity (% volume)	Saturation coefficient	Microporosity (% saturation)	Capillarity coefficients French A ($g.cm^{-2}.min^{-1/2}$)	Belgian GC	Belgian G	Crystallisation test (% weight loss)
Blue rag							
(3800)	23.4	0.62	76	2.6	−1.2	0.9	0, 0
3600	22.2	0.63	81	(1.5)	(−0.6)	(10.0)	0, 0
Blue stone							
(3300)	26.1	0.66	65	1.3	4.9	68.8	2, 1
3600	26.1	0.72	83	(1.9)	(0.4)	(13.5)	0, 0
Brown stone							
(2600)	33.1	0.76	76	5.0	1.2	7.8	9, 7
2900	35.1	0.74	70	(4.0)	(−0.04)	4.1	1, 1

Durability
Blue rag: Class A
Blue stone: Class A
Brown stone: Class B to C

HOVINGHAM STONE

Grid reference SE 669 753

Owner and quarry location Mrs E A Ashbridge, The Stoneworks, Hovingham, N Yorks. Tel: Hovingham (065-382) 200

The quarry is at Hovingham itself and is just off the B1257, the road to Malton. There are old quarries nearby and further areas for exploitation.

Petrography *(Plate 5)*
Hovingham Stone is an oolitic limestone of late Jurassic age. Two stones are available, under about 450 mm of overburden; one is creamy coloured whilst the other is blue-grey. Both contain small shell fragments. Blocks of 750 mm on bed are commonly available up to a maximum of 900 mm.

Reference buildings
Hovingham Stone is generally used as a walling stone. The following are some examples: housing at the old quarry entrance in Hovingham itself; various architect-designed individual houses including one in Gilling East (on the B1363) on the main street near the church and another in Ryegate (street) Helmsley in the North Yorkshire National Park.

In 1980 the stone was used for an extension to the 'Peveril-of-the-Peak' Hotel at Ashbourne, Derbyshire. It was used for flooring at Hunslet Methodist Church near Leeds, also in 1980.

Results of tests

Velocity of sound ($m.s^{-1}$)	Porosity (% volume)	Saturation coefficient	Microporosity (% saturation)	Capillarity coefficients French A ($g.cm^{-2}.min^{-1/2}$)	Belgian GC	Belgian G	Crystallisation test (% weight loss)
White stone							
2800	16.6	0.94	81	(4.0)	(3.9)	(18.5)	59, 59
3000	15.5	0.95	85	(5.0)	(3.8)	(18.9)	40, 53
Blue stone							
4100	11.5	0.86	88	(1.6)	(3.7)	(18.0)	5, 8
3900	12.0	0.92	89	(2.0)	(3.9)	(19.1)	17, 15

Durability
White stone: Class E
Blue stone: Class C

KENTISH RAGSTONE

Grid reference TQ 655 571

Owners Amey Roadstone Corpn Ltd, SE Division, Basted House, Borough Green, Kent. Tel: West Malling (0732) 884242

Quarry location Offham, Nr West Malling, Kent. Tel: West Malling (0732) 843071

Kentish Ragstone is obtained from the Offham Quarry which is 1 mile west of West Malling. The latter is on the A228 in Kent. There are numerous old quarries nearby and further areas to exploit.

Petrography *(Plate 6)*

Kentish Ragstone is a glauconitic sandy limestone from the Lower Greensands of early Cretaceous age. Glauconite is a complex mineral of potassium and iron aluminium silicate. A predominance of silica indicates a sandstone and in fact the amount of silica of some beds rises to the level where these beds are indeed regarded as sandstones and not limestones.

The stone with which we are concerned is the hard blue-grey limestone often with small speckles of green (due to the glauconite). It is obtained from a 27 m face and is interlaced with a soft crumbly stone. The latter is more like a sandstone with a high silica content. The average depth of limestone on bed is 230 mm, the maximum is 600 mm.

The stone is known as a ragstone because it is hard and rubbly. This is in contrast to a freestone which can be easily worked in any direction.

Reference buildings

Because it is so difficult to work, Kentish Ragstone is generally found as uncoursed walling with a different stone providing the dressing. However, owing to a shortage of good stone it was extensively used in and around Kent as dressed block and even sometimes as window tracery.

The Romans were the first to use the stone in London for building walls. Many churches in the Middle Ages were built in the stone and the White Tower in London is of Kentish Ragstone with a French stone used for the dressings. Stone from the present quarry was used in 1981 to build garden walls at the Tower. These walls are not on the river side of the Tower and can be reached by turning left after crossing Tower Bridge.

Rochester Castle (1078) was built chiefly in this stone. Old Soar Manor near Plaxtol, just to the south-west of the present quarry, and of course many Kent churches were all built in Kentish Ragstone. Knole House, near Sevenoaks, is largely constructed with this stone. Here, in one portion of the house, the stone was worked successfully to form some of the dressings. In 1981 stone from the present quarry was used for restoration work at Knole. About 1850 it was used at the City Prison, Holloway. Kentish Ragstone can also be seen in Maidstone Bridge completed in 1981.

(continued)

Results of tests   Kentish Ragstone

Velocity of sound (m.s^{-1})	Porosity (% volume)	Saturation coefficient	Microporosity (% saturation)	Capillarity coefficients French A (g.cm^{-2}.min$^{-1/2}$)	Belgian GC	Belgian G	Crystallisation test (% weight loss)
(4200)	2.7	0.92	57	0.2	2.1	−9.0	0, 0
4300	3.1	0.83	64	(0.2)	(6.2)	(82.0)	0, 0

Durability
Class A

Although the results from the tests show Class A stone, past experience and the levels of silica of some beds indicate Class B.

KETTON STONE

Grid reference SK 985 055

Owners The Ketton Portland Cement Co Ltd, Albion Works, Sheffield, S4 7UL. Tel: Sheffield (0742) 26311

Quarry location Ketton, Nr Stamford, Lincolnshire. Tel: Stamford (0780) 720501

The quarry is on the A6121 which is off the A1 near Stamford. This extremely large quarry (the total area is 1820 hectares), with its cement works, can be seen on the left whilst travelling northwards on the A1, and in fact most of the output is crushed for cement. The stone has been quarried for centuries, possibly as far back as Roman times. There are huge reserves of stone.

Petrography *(Plate 6)*
Ketton Stone is an oolitic limestone from the Lincolnshire Limestone formation of middle Jurassic age. Ketton is the classic oolitic limestone in terms of appearance in that it seems to consist entirely of ooliths. However, the ooliths are joined together by small bridges of calcium carbonate.

At the time of the visit the freestone was being extracted at three locations around this vast quarry. It is found under 3 m of rag which is unsuitable for building purposes. There is 1.8 m of freestone giving blocks, on bed, of 900 to 1200 mm. In some parts of the quarry the freestone occurs in columns as opposed to the more normal bedding pattern of lying under the rag.

The blocks of stone are blasted and then pushed down to the floor of the quarry from above. However the quarrymen stated that Ketton is not adversely affected by the blasting. The stone is known as 'horses head' because of the random shapes and this leads to some wastage.

Ketton is generally described in older books as an even-textured stone ranging in colour from cream to brownish buff and from pale to a warm pink. The stone currently extracted shows more colour variation with some pronounced bedding and white veins. It is 'plummy', ie it has some hard shiny bits. A firm of stone masons – Rattee and Kett of Cambridge – with much experience in the use of Ketton, supplied some previously extracted Ketton from their store yard for comparison with the currently available stone. The results for both the 'stored' and 'new' stone are given, together with the results of a sample which was mainly white in colour.

Reference buildings
In the early 1800s Ketton Stone was used for the Bottle Lodges at Burghley House and later, in 1828, for the centre-piece and dressings of Stamford Hospital. Of course, Ketton has been extensively used in and around Cambridge, eg at Downing, Trinity, and for the front gateway dressings at St Catharine's College. It may be seen further afield at Audley End House, near Saffron Waldon, Essex, and the buildings on the west side of Westminster Hall, Palace of Westminster.

For many years in this century most of the output from the quarry was used for cement making. Any building stone was held by Rattee and Kett and was mostly used for restoration work, an example of which can be seen at the North Stable Block at Woburn Abbey, Bedfordshire. In the early 1970s the stone was used for pilasters, cornices, mullions and to encase two columns facing the main house. However, a new seam of stone, suitable for building purposes, has been discovered, and while most of this is also held by Rattee and Kett, the quarry will consider direct enquiries. Examples of the use of the new seam are as follows:

Marlborough School, Bedford — copings on north gable, 1981
King's College Chapel, Cambridge, Wilkins Screen — two bays are being renewed each year over a 5-year period which started in 1981
Grimsthorpe Castle, near Bourne in Lincolnshire — new chimney caps, 1981
Burghley House, Stamford — new chimney caps, 1982
Tower Hospital, Ely — new mullions, 1982

Results of tests   Ketton Stone

Velocity of sound (m.s^{-1})	Porosity (% volume)	Saturation coefficient	Microporosity (% saturation)	Capillarity coefficients French A (g.cm^{-2}.min$^{-1/2}$)	Belgian GC	Belgian G	Crystallisation test (% weight loss)
Old stone							
2700	23.5	0.64	42	(13.9)	(−3.0)	(−7.4)	4, 2
(4200)	24.0	0.65	43	4.5	−2.3	−7.7	1, 0
New stone							
2800	23.4	0.67	43	(5.0)	(−1.7)	(−8.2)	0, 0
(2700)	25.1	0.64	39	10.1	−2.3	−8.8	3, 5
White stone, new piece							
2900	24.8	0.65	41	(7.7)	(−2.4)	(−7.4)	0, 0

Durability
Old stone: Class B
New stone: Class B
White stone, new piece: Class A

KILKENNY MARBLE

Owners Feely and Sons Ltd, Boyle, Co Roscommon, Eire.
Tel: Boyle 66 (via operator)

Quarry location Kellymount Quarries, Paulstown, Co Kilkenny, Eire.
Tel: Ballon (0503) 26191 (via operator)

This quarry is 0.75 mile north of Whitehall on the N9. There is a large area for future exploitation.

Petrography *(Plate 6)*
Kilkenny Marble is of early Carboniferous age. Under up to 10 m of overburden there are seven beds of stone which are all uniform in quality. However, only five beds of the 8.4 m face are worked currently. One of these beds is so deep that it provides stone of 3.3 m on bed whilst the other beds give 900 mm blocks. The stone is dense and smooth and is dark blue-grey in colour; small pieces of shell are evident.

Reference buildings
This stone is usually known as Kilkenny Marble even though it is not a marble in the geological sense. It is called a 'marble' because it takes a good polish. It is currently being used for restoration at Wells Cathedral to provide the slender blue-grey columns which support the canopies over the statues of the West Front.

It was used for flooring at Coventry Cathedral and for restoration work at Galway Cathedral in 1966. It was used for cladding and flooring at University College Dublin in 1972–73.

Results of tests

Velocity of sound (m.s^{-1})	Porosity (% volume)	Saturation coefficient	Microporosity (% saturation)	Capillarity coefficients French A (g.cm^{-2}.min$^{-1/2}$)	Belgian GC	Belgian G	Crystallisation test (% weight loss)
5500	0.7	0.98	18	(0.06)	(1.4)	(12.2)	0, 0
(5300)	0.6	0.67	11	0.05	−3.2	−19.2	0, 0

Durability
Class A

LECARROW STONE

Owners James Murphy and Sons Ltd, Murphystown Road, Sandyford, Co Dublin, Eire. Tel: Dublin (0001) 985006

Quarry location Lecarrow, Co Roscommon, Eire.
Tel: Roscommon (0903) 7131 (via operator)

The quarry is at Lecarrow itself on the Roscommon to Athlone road, the N61. There are old quarries nearby and large areas for future exploitation.

Petrography *(Plate 6)*
Lecarrow Stone is of early Carboniferous age. There are seven beds which make up a 13.5 m deep face of stone. The depth of stone on bed varies from 1.2 m up to 2.4 or 2.7 m. The stone is a dark blue-grey colour and is dense and smooth; numerous pieces of shell are evident.

Reference buildings
Lecarrow Stone is used for a variety of applications ranging from walling to cladding and dressings. The following are some examples:

Parnell Bridge, Cork – 1975
Bridge at Carrick on Suir – 1970s
Belsield College, Dublin University – facing and paving, 1977
Bank of Ireland, Roscommon – lintels and mullions, 1980
Church at Ashbourne, Co Meath – lintels and mullions, 1981
Christchurch Cathedral, Dublin – restoration of dressings, 1982

Results of tests

Velocity of sound (m.s^{-1})	Porosity (% volume)	Saturation coefficient	Microporosity (% saturation)	Capillarity coefficients French A (g.cm^{-2}.min$^{-1/2}$)	Belgian GC	Belgian G	Crystallisation test (% weight loss)
6100	0.4	0.98	29	(0.2)	(3.5)	(59.9)	0, 0
(6000)	0.5	0.66	10	0.02	−5.9	−36.6	0, 0

Durability
Class A

MANDALE STONE

Grid reference SK 175 688 (for Sheldon)

Agents Bolehill Quarries Ltd, Wingerworth, Chesterfield, Derbyshire.
Tel: Chesterfield (0246) 70244

Quarry location Nr Sheldon, Derbyshire

This quarry, known as the 'Once-a-week Quarry', is on the south side of the road from Sheldon to Flagg, approximately midway between Buxton and Bakewell off the A6. It is extremely difficult to find and would-be callers should contact Mr Sedgewick at the above address and he will escort them to the quarry. There are large areas for future exploitation. The quarry was reopened in 1977.

Petrography *(Plate 6)*
Mandale Stone is of early Carboniferous age. It is a dense, light grey stone. There are several beds providing stone from 150 to 230 mm up to about 1 m on bed. Large blocks are obtainable.

Reference buildings
Sawn Mandale Stone was used for the extension to St Mary's Hospital, Bakewell, in 1978. It was used as 100 mm cladding to stairwells leading out of the moat gardens onto Tower Hill at the Tower of London in 1980. Also in 1980 it was used to provide new triangular-shaped coping stones for walling at the Ulverston Diversion, Cumbria, at New Church Lane. However, Mandale is perhaps most used for random walling. An example of this can be seen at the walls either side of the two new bridges at Ashford in the Water near Bakewell, built in 1980.

Results of tests

Velocity of sound (m.s^{-1})	Porosity (% volume)	Saturation coefficient	Microporosity (% saturation)	Capillarity coefficients French A (g.cm^{-2}.min$^{-1/2}$)	Belgian GC	Belgian G	Crystallisation test (% weight loss)
5800	2.7	0.74	59	(0.2)	(−1.4)	(−24.7)	0, 0
5900	1.9	0.84	39	(0.1)	(−1.1)	(−27.5)	0, 0

Durability
Class A

NASH ROCKS

Grid reference SO 243 578

Owners Nash Rocks Stone Lime Co Ltd, PO Box 1, Kington, Herefordshire, HR5 3LQ. Tel: Kington (0544) 230711

Quarry location Nr Kington, Herefordshire

There are two adjacent quarries — Strinds and Dolyhir — 3 miles west of Kington on the A44. They are large quarries with good reserves of stone. The nearby quarry actually known as Nash Quarry is no longer worked.

Petrography *(Plate 6)*
Both quarries provide a stone of Silurian age. The face at the Strinds quarry is 31.5 m in depth and there is thought to be a further 12 m below this level. At Dolyhir the face is 36 m. Both quarries provide blocks which are about 450 mm on bed. The stone is dark grey and mottled with lighter grey patches.

Reference buildings
Although the bulk of the output from these quarries is used for aggregates, the stone has been used locally for housing, agricultural buildings and bridgeworks. The stone is used for retaining walls on roads, etc. Here it is often put into wire baskets known as gabions and used to hold back banks. Gabions of Nash Rocks were used at the bridge in Kington.

Results of tests

Velocity of sound (m.s^{-1})	Porosity (% volume)	Saturation coefficient	Microporosity (% saturation)	Capillarity coefficients French A (g.cm^{-2}.min$^{-1/2}$)	Belgian GC	Belgian G	Crystallisation test (% weight loss)
Strinds Quarry							
5600	0.5	0.72	45	–*	–*	–*	0, 0
5600	0.9	0.54	40	(0.03)	(–17.5)	(–48.9)	0, 0
Dolyhir Quarry							
4400	1.0	0.83	36	–*	–*	–*	0, 0
4200	1.0	0.68	26	–*	–*	–*	0, 0

*Sample was not a suitable shape for this test

Durability
Strinds Quarry: Class A
Dolyhir Quarry: Class A

NAVAN STONE

Owner and quarry location Mr D Acheson, Navan Quarry, Armagh, Co Antrim, N Ireland. Tel: Armagh (0861) 522667

This large quarry lies 2 miles west of Armagh on the A28. There are other quarries nearby and further areas to exploit. The stone is not normally quarried for building purposes but specific orders are accepted. It is normally blasted out of the quarry and crushed for aggregates.

Petrography *(Plate 6)*
Navan Stone is of early Carboniferous age. Stone is extracted from two faces each of 30 m giving slight colour variations. The sample tested was a light grey colour with pale pink tinges; shell fragments were present.

Reference buildings
Local builders select, for themselves, any stone they require for housing and fireplaces.

Results of tests

Velocity of sound ($m.s^{-1}$)	Porosity (% volume)	Saturation coefficient	Microporosity (% saturation)	Capillarity coefficients French A ($g.cm^{-2}.min^{-1/2}$)	Belgian GC	Belgian G	Crystallisation test (% weight loss)
4200	1.0	1.00	41	(0.08)	(−2.2)	(15.9)	0, 0
4700	1.0	0.93	50	(0.06)	(−2.5)	(−1.3)	0, 0

Durability
Class A

ORTON SCAR STONE

Grid reference NY 641 103

Owners Cumbria Stone Quarries Ltd, Silver Street, Crosby Ravensworth, Cumbria, CA10 3SA. Tel: Ravensworth (093-15) 227

Quarry location Nr Orton, Cumbria

The quarry is to the north-east of the village of Orton off the B6260 and to the east of Orton Scar. Orton is just north of Junction 38 on the M6. There are areas nearby for future work.

Petrography *(Plate 6)*
Orton Scar Stone is of early Carboniferous age. There are three beds of stone under approximately 1.2 m of overburden. The top bed is thin and not always usable. The other two beds are similar to each other. There is some colour variation in the stone and the sample tested was light grey in colour. The average depth of stone on bed is 230 mm.

The Orton Scar and Salterwath quarries are both worked by men from Orton.

Reference buildings
Orton Scar Stone, together with Hopton Wood, has been used for flooring at Birmingham Cathedral. In approximately 1965 it was used for flooring at Abbot Hall in Kendal. It was also used in County College at Lancaster University in 1971 for cladding and County Hall in Preston in 1973 for cladding and internal walls. In 1980 it was used at the Trustee Savings Bank in Kendal. Here the walling is pitch-faced Orton Scar with rubbed stone for the sills, steps and logo. It is used for house cladding.

Orton Scar has also been used for monumental works for the last 150 years. It will accept a good polish. The stone has been used extensively by the Commonwealth War Graves Commission as far apart as Tunisia and Iceland.

Results of tests

Velocity of sound ($m.s^{-1}$)	Porosity (% volume)	Saturation coefficient	Microporosity (% saturation)	French A ($g.cm^{-2}.min^{-1/2}$)	Belgian GC	Belgian G	Crystallisation test (% weight loss)
6000	2.9	0.69	56	(0.2)	(−0.9)	(−3.2)	0, 0
5600	4.4	0.72	58	(0.2)	(−2.1)	(7.0)	0, 4

Durability
Class A

PENMON STONE

Grid reference SH 635 813

Owners The Steetley Co Ltd, Gateford Hill, PO Box 6, Worksop, S81 8AF.
Tel: Worksop (0909) 4551

Quarry location Penmon, Nr Beaumaris, Anglesey, N Wales.
Tel: Beaumaris (0248) 78201

The village of Penmon is off the B5109 about 6 miles north of Beaumaris. The quarry is past the village, very near to the north shore of the point. There are two separate faces which are about 300 m apart. From one a black limestone is extracted, whilst the other, which contains a white limestone, was not being worked at the time of the survey visit. Thus, only the black stone was tested although there are large reserves of both types of stone.

Petrography *(Plate 6)*
Black Penmon Stone is of early Carboniferous age. It is a dark, shiny limestone with no obvious grain and is, therefore, smooth in appearance. Some blocks have small white veins running through them. The largest available block is 600 mm square but most blocks are much smaller. The stone is blasted from a 9 m vertical cliff face where it is layered in 150 to 200 mm blocks creating a step-like appearance.

Reference buildings
The black limestone was used for decorative slabs and paving in medieval times. Today it is used locally by the council to build retaining walls, and for road making and aggregates.

The white limestone was used in the Menai Bridge and in a dovecot built near the quarry in 1608. This stone was also used for lime-burning in nearby kilns.

Since the time of the survey visit the quarry has been taken over. It is understood that the new owners intend to extract the white stone again.

Results of tests

Velocity of sound (m.s^{-1})	Porosity (% volume)	Saturation coefficient	Microporosity (% saturation)	Capillarity coefficients French A (g.cm^{-2}.min$^{-1/2}$)	Belgian GC	Belgian G	Crystallisation test (% weight loss)
2200	0.5	0.87	67	–*	–*	–*	0, 0
2200	1.6	0.77	72	–*	–*	–*	0, 12

* Sample was not a suitable shape for this test

Durability
Class B

PORTLAND STONE

Like most famous quarries those on the island of Portland were known to the Romans. The stone was worked intermittently throughout the Middle Ages. In the early fourteenth century some stone was sent to Exeter for use at the cathedral and some to London for the Palace of Westminster. But it was not until the seventeenth century, with the introduction of frame-saws and water power, that cutting so hard a stone became economic on a large scale.

James I appointed Inigo Jones chief architect and surveyor-general. Under this appointment, Jones was required to survey the Crown lands on the island of Portland and it was he who first brought the stone, in quantity, to London. In 1619 he used Portland to build the Great Banqueting Hall of Whitehall Palace and later for restoration work at old St Paul's Cathedral.

After the Great Fire in 1666 it was, of course, Sir Christopher Wren who ensured Portland's success in London. The new St Paul's Cathedral and most of the new city churches were built or faced in Portland Stone and indeed Wren used over a million tons of the stone in London alone.

Once the initial rebuilding of London was over, Portland continued to be specified for many major buildings although demand was not as great as during the seventeenth century. The British Museum (1753), Somerset House (1776 – 92), The Bank of England (1826) and County Hall (1933) are just some of the magnificent Portland buildings.

During the eighteenth century Portland Stone first appeared in Cambridge. It was used for the Fellows' Building at King's College (1723 – 29) and in the Senate House (1722 – 30) both built by Gibbs. The following century it was used for the Old University Library and for the Fitzwilliam Museum.

Apart from Rufus Castle (rebuilt in the fifteenth century) there are no large monuments of the local stone on the island of Portland. There is no 'home market' for Portland in the sense that Bath Stone was used for the city of Bath and the Lincolnshire Limestones were used at Stamford, Grantham, etc. The stone's reputation has been made away from Dorset, largely in London.

Weathered Portland Stone buildings are always distinctive in appearance because the surfaces exposed to rain become very white whilst those which are protected accumulate dirt and turn black. Thus a Portland building stands out because of its characteristic black and white appearance. It follows, therefore, that Portland Stone does not look well at every location particularly if there are other stone buildings which blend in with their surroundings. For example, there is little Portland in Oxford – the columns and pilasters of the Ashmolean Museum (1840s), the cupola over the High Street gateway at The Queen's College (1909) and Christ Church Library, refaced in 1962, are some exceptions. In addition the Ashmolean now has a Portland roach plinth. On the other hand, most of the principal Georgian buildings in Dublin are partly or completely faced in Portland and they contrast well with the other, predominantly brick, buildings. Portland looks particularly well when used for the dressings of brick buildings. The Fountain Court at Hampton Court Palace, built by Wren, is one example of the use of Portland dressings.

There are three famous beds of Portland Stone – roach, whit bed and base bed. Huge blocks of stone are available from each bed. The stone is found under a great deal of overburden, perhaps as much as 9 m. Within the overburden there may be some stone from the Purbeck Beds. This stone is all discarded, in contrast to the recognised Purbeck quarries on the mainland where every bed of stone is put to some use.

The roach is an extremely shelly stone with a large number of holes scattered throughout it. The holes are due to the removal of fossil shells by percolating rain. The finer-grained part of the stone is very similar to whit bed. Roach is very durable and although previously little-used is employed nowadays as a decorative cladding. Only one of the preceding examples of use of Portland Stone is roach, all the rest are whit or base bed.

The whit bed is a fine-grained stone containing shell fragments and it lies below the roach; it too is very durable, although not quite as durable as roach.

Between the whit and base beds, the last of the three recognised beds, there is often a band of poor stone known as curf. This is generally regarded as waste and must be removed before the base bed can be extracted. The base bed, also known formerly as best or bottom bed, is softer than whit and not thought to be as durable. It is not so obviously oolitic in appearance as the whit bed but contains less shell and is suitable for carving.

Apart from the advantages of being fine-grained, durable and available in large blocks, Portland also weathers very evenly.

Four quarries were visited at the time of the survey. Three of these are owned by the same company — Kingston Minerals. Since the visits two further quarries have been reopened by another company who also provided stone for testing.

PORTLAND STONE – BOWYERS QUARRY

Grid reference SY 686 725

Owners Albion Stonemasonry Ltd, Boundary Road, Merton, London SW19.
Tel: 01-542 1045

Quarry location Bowyers Quarry, Wide Street, Easton, Portland

This quarry is off Wide Street in Easton. There are plentiful reserves of stone.

This quarry was not open at the time of the visits but samples were obtained for inclusion in the testing programme.

Petrography *(Plate 7)*
Portland Stone is an oolitic limestone from the Portland Beds of late Jurassic age. The beds of stone are as follows:

Roach – this bed is 1.5 m deep.

Whit – greyish-white fine-grained stone in a 1.2 m deep bed.

Curf – waste stone, 2.4 m deep, which is removed to reveal the base bed.

Base – greyish-white fine-grained stone, not as oolitic in appearance as the whit bed. The depth of this bed varies from 1.2 to 1.5m.

Both whit and base beds are available in blocks of 1.2 m on bed.

Reference buildings
As this quarry has only been opened a short time there are few locations where the stone has been used. The output from the quarry is taken to the head office at Merton where there is a masonry yard. To date whit bed has been used at the Cutlers Gardens office complex near Liverpool Street Station in London. Arches, plinths, band courses (string courses with no moulding) and cornices are all in Portland Stone. This job was completed in 1981. Stone from the Kingston Minerals' quarries was used for the earlier work on this contract before Bowyers Quarry became operational.

The stone was also used for new sills and string courses at the Red Lion Pub, Whitehall, in 1981.

Results of tests

Velocity of sound ($m.s^{-1}$)	Porosity (% volume)	Saturation coefficient	Microporosity (% saturation)	French A ($g.cm^{-2}.min^{-1/2}$)	Belgian GC	Belgian G	Crystallisation test (% weight loss)
Whit bed							
3900	13.7	0.87	43	(8.0)	(1.9)	(26.0)	18, 17
(3800)	13.4	0.89	45	6.8	2.3	26.2	19, 17
Base bed							
4000	13.0	0.96	70	(3.0)	(4.1)	(30.8)	78, 68
(3900)	12.9	0.97	75	3.2	4.3	32.9	73, 75

Durability
Whit bed: Class D
Base bed: Class E

PORTLAND STONE – INDEPENDENCE QUARRY

Grid reference SY 692 718 (centre of Easton)

Owners Albion Stonemasonry Ltd, Boundary Road, Merton, London SW19.
Tel: 01-542 1045

Quarry location Independence Quarry, Grove Road, Easton, Portland

The entrance to this quarry is in Grove Road and it lies behind the football ground. Work started at the end of 1982 and there are plenty of reserves of stone. The quarry was opened too late to be included in the full testing programme and only a crystallisation test has been carried out. The quarry was formerly known as the Convicts Quarry.

Petrography
Portland Stone is an oolitic limestone from the Portland Beds of late Jurassic age. The beds are as follows but the depths may change as work progresses.

Roach – 600 mm deep.

Whit – two beds, each 1.2 m deep.

Middle – 1.2 m deep of a stone which is neither base bed nor whit bed but is a distinctive workable bed.

Base – 1.2 m deep.

Reference buildings
It is too early for the stone to have been used in any major contracts.

Results of tests

| Crystallisation test (% weight loss) |||
Roach bed	Whit bed	Base bed
56, 49, 17	19, 21, 10	35, 39, 35
29, 56, 16	24, 12, 18	31, 36, 41
31, 15	11, 14	33, 37

Durability
Roach bed: Class D
Whit bed: Class C to D
Base bed: Class D

PORTLAND STONE – KINGSTON MINERALS' QUARRIES

Grid reference SY 695 716 (Bumpers Lane)

Owners Kingston Minerals Ltd, Bumpers Lane, Portland. Tel: Portland (0305) 820207

Kingston Minerals own several quarries on the island of Portland. At the time of the survey two active quarries – Coombefield and Weston – were visited. In addition, the Fancy Beach Quarry was visited and although this was not being actively worked, and there are currently no plans to do so, there are vast stores of stone from previous workings at this location. Samples of stone from all three quarries were obtained for testing.

Anyone wishing to visit the quarries should call in at the head office which is off Wakeham Street in the village of Easton.

Petrography *(Plate 7)*
All the quarries are working the oolitic Portland Stone from the Portland Beds of late Jurassic age. The sequence of beds at each quarry is as follows:

Coombefield Quarry (Grid reference SY 689 703)

Roach — 1.8 to 2.1 m deep bed of typical buff coloured shelly stone with numerous holes.

Whit — 2.4 m deep bed.

Base — not worked at this quarry owing to flaws.

A sample of roach was obtained for testing.

Fancy Beach Quarry (Grid reference SY 689 726)

Roach – 3.0 to 3.6 m deep bed of a buff coloured shelly stone with numerous small round and deep holes.

Whit – 1.2 to 1.8 m deep bed of a buff coloured shelly stone.

Base – 1.8 m deep bed.

Samples of roach and whit were obtained for testing.

The samples from Fancy Beach were noticeably darker than other Portland samples. Kingston Minerals can supply all three stones at depths on bed of between 450 and 900 mm.

Weston Quarry (Grid reference SY 689 709)

Roach – 3.0 to 3.6 m deep bed.

Whit – 1.2 m deep bed of a creamy coloured stone.

Base – the top metre of this bed is discarded leaving usable stone of 1.2 to 1.8 m deep. The stone is creamy coloured and finer grained than the whit but regarded, by the quarry, as not suitable for carving.

Samples of whit and base beds were obtained for testing.

Reference buildings
The fame of the Portland quarries is such that many towns all over the country can boast a major building completed in the stone. Of course, Portland Stone is still used extensively in London. The following are just a few of the more recent contracts involving the stone:

Royal Festival Hall, London — ashlar, 1962
Grosvenor House Hotel, Sheffield — ashlar, 1964
Bradford Library — ashlar, 1964
Blackwall Tunnel, London — ashlar, 1965
Mersey Tunnel, Birkenhead — ashlar, 1967
Stoke-on-Trent Library — ashlar, 1968
Sun Alliance Building, Horsham — ashlar, 1970
Slough Library — ashlar, 1972
Queen Anne's Mansions, London — cladding, 1972
Intercontinental Hotel, Park Lane, London — cladding, 1976
AUEW Headquarters, London — cladding, 1977
Kenstead Hall, London — balustrading, moulded steps and plinths, 1979
Woolworth, Cheapside, London — cladding, 1980
St Anne's Cathedral, Belfast — restoration of moulding carried out over a number of years and completed in 1979
Devonshire House, Piccadilly, London — repair to cladding, 1981
Gateway Building Society, Worthing — extension to existing building, 1981
Maidstone Crown Court — cladding, 1982

(continued)

Results of tests Portland Stone – Kingston Minerals' Quarries

Velocity of sound (m.s^{-1})	Porosity (% volume)	Saturation coefficient	Microporosity (% saturation)	Capillarity coefficients French A (g.cm^{-2}.min$^{-1/2}$)	Belgian GC	Belgian G	Crystallisation test (% weight loss)
Coombefield Quarry – Roach bed							
(4100)	15.3	0.62	52	3.1	−3.4	−12.1	5, 9
4400	15.7	0.61	–	(0.9)	(−4.2)	(−23.7)	–
4100	18.2	0.56	37	(1.8)	(−5.1)	(−22.1)	–
(3800)	17.3	0.60	–	1.6	−3.3	−17.3	6, 6
Fancy Beach Quarry – Roach bed							
3700	21.6	0.56	–	(2.4)	(−4.1)	(−17.2)	–
(3600)	23.1	0.55	12	8.3	−5.4	−18.4	0, 0
3800	22.9	0.61	–	(8.3)	(−3.9)	(−10.7)	0, 0
(3900)	22.9	0.59	14	11.7	−5.8	−19.9	–
Fancy Beach Quarry – Whit bed							
3900	22.3	0.58	23	(2.4)	(−2.9)	(−11.3)	0, 0
(3900)	23.4	0.59	–	12.9	−4.3	−7.3	–
(4000)	22.4	0.57	43	7.9	−4.0	−6.1	1, 2
3700	22.8	0.58	–	(5.3)	(−3.5)	(−12.5)	–
Weston Quarry – Whit bed							
3400	22.7	0.66	40	(9.4)	(−1.5)	(−0.2)	12, 13
(3900)	21.6	0.65	–	11.0	−2.7	−4.5	20, 27
(3900)	21.3	0.66	27	3.0	−1.2	−1.3	–
3500	20.9	0.67	–	(9.9)	(−2.5)	(−1.0)	–
Weston Quarry – Base bed							
(3800)	19.1	0.70	57	5.2	−1.0	4.6	–
4000	18.5	0.70	–	(5.8)	(−1.2)	(2.9)	23, 23
(3700)	20.9	0.67	–	3.8	−1.1	0.7	–
4000	17.8	0.71	43	(5.7)	(1.0)	(4.9)	28, 30

– Not tested

Durability

Coombefield Quarry – Roach bed: Class B
Fancy Beach Quarry – Roach bed: Class A
                             Whit bed: Class A
Weston Quarry – Whit bed: Class C to D
                     Base bed: Class D

PORTLAND STONE – SHEAT QUARRY

Grid reference SY 690 697

Owners Portland Stone Products Ltd, Weston Quarries, Portland.
Tel: Tarrant Hinton (025-889) 244

Quarry location Sheat, Southwell, Portland. Tel: Portland (0305) 821226

This quarry is just off the road to Portland Bill from Southwell very near the village of Southwell. The quarry was opened around 1970 and although there is the potential to expand to adjoining land this quarry was to be temporarily closed at the end of 1982.

Petrography *(Plate 7)*
This is typical of the other quarries on Portland in that the stone is an oolite from the Portland Beds of late Jurassic age. However, it is atypical in that there appears to be little difference between the beds of stone. The sequence is as follows:

Whit bed – the total depth of this bed is about 2.7 m. Within it three different layers known locally as tiers are identified. The top tier is about 1 m deep of creamy-buff stone with quite a few pieces of shell together with a few small holes. Then there is the middle tier which is also shelly but more coarse than the top tier and is 680 mm deep. The final tier is the seedy tier which is just over 900 mm deep and is also shelly and coarse.

Base bed – again many shell pieces are evident and although this is generally fine grained there are some coarse patches. The bed is 600 mm deep.

Roach – coarse stone with shell fragments but without the holes of typical Portland roach. The bed is 750 to 900 mm deep. Note that in other Portland quarries the roach is the top bed.

Below the roach there is a further base bed 2.1 to 2.4 m deep but this is not quarried. Although the depth of stone on bed is not great (on average 680 mm) exceptionally long blocks can be obtained from this quarry.

Reference buildings
Stone from this quarry was used for the base of the Polish Katyn Memorial which is in Gunnersbury Park Cemetery in London and was built in the late 1970s. The Barclay's Bank eagle for their new branch in Gracechurch Street, London, was carved in Portland in 1981. At the Burrel Museum in Glasgow the stone has been used (1983) for a plinth below the glazed north edge of the building. Internally it has been used for paving, balustrades, skirting and above the show cases. In 1981 – 82 the stone was used for restoration work at Tower Bridge, London.

(continued)

Results of tests   Portland Stone – Sheat Quarry

Velocity of sound (m.s⁻¹)	Porosity (% volume)	Saturation coefficient	Microporosity (% saturation)	Capillarity coefficients French A (g.cm⁻².min⁻¹ᐟ²)	Belgian GC	Belgian G	Crystallisation test (% weight loss)
Top tier							
3900	20.4	0.61	27	(8.8)	(−3.2)	(−6.6)	12, 9
3900	19.9	0.62	34	(7.3)	(−2.7)	(−6.3)	0, 4
Middle tier							
3500	20.8	0.59	26	(6.9)	(−2.7)	(−7.4)	0, 0
3600	20.3	0.59	28	(10.1)	(−3.5)	(−7.5)	3, 2
Seedy tier							
3700	17.7	0.68	49	(5.0)	(−1.5)	(3.9)	25, 19
3700	17.5	0.68	45	(3.2)	(−0.9)	(4.5)	17, 18
Base bed							
4100	16.4	0.68	58	(3.0)	(−0.8)	(4.3)	22, 22
4100	16.6	0.66	59	(2.9)	(−1.2)	(2.2)	18, 19
Roach bed							
3400	20.8	0.64	39	(10.8)	(−2.8)	(−1.0)	22, 17
3200	21.2	0.64	–	(8.8)	(−2.2)	(−0.8)	12, 18

– Not tested

Durability
Top tier: Class B to C
Middle tier: Class A to B
Seedy tier: Class D
Base bed: Class D
Roach bed: Class D

PURBECK STONE

There are numerous quarries in the Isle of Purbeck in Dorset. The quarrying and working of Purbeck Stone has traditionally been carried on by local families. The skills and the quarries are passed down from father to son — most of the quarries have been worked by the same families for centuries. Three types of stone are available:

Purbeck Limestone of which there are over 20 different beds,

Purbeck Marble, not a marble in the geological sense but a hard shelly limestone which takes a good polish, and

Purbeck-Portland Stone which is similar to Portland Stone in that it is a grey-white fine-textured stone but is much harder to work. This stone is only found at St Aldhelm's Head.

The Purbeck Beds can be split into Upper, Middle and Lower divisions. The Upper division was laid down under fresh-water conditions whilst the Middle and some of the Lower division were laid down under estuarine and marine conditions. Towards the top of the Upper division lies the Purbeck Marble. The main Purbeck Limestone lies in the Middle division and the Purbeck-Portland Stone lies in the Portland Beds underlying the Lower division of Jurassic age.

Six quarries were visited at the time of the survey and reports and results for these quarries follow in the succeeding pages. In addition to these six, there are, according to *Stone Industries* Vol 15 No 7, a further five quarries. These are:

Wellman and Harris, Halcyon Cottage, The Hyde, Langton Matravers.
    Tel: Swanage (092-92) 4188

D and P Lovell Quarries, 2a Cranborne Road, Swanage. Tel: Swanage (092-92) 2657

Lewis and Sons, 24 Park Road, Swanage. Tel: Swanage (092-92) 2114

H F Bonfield and Sons, Springdale, Gully Combe, Langton Matravers.
    Tel: Swanage (092-92) 3697

Cobb Bros, Eastington Road, Worth Matravers.

The *Natural stone directory*[1] for 1980–81 also lists G Hancock and Sons of 117a High Street, Swanage. The premises in the High Street are a masonry yard and they are working stone from Langton Matravers. Reference to the latest edition of the *Natural stone directory* will provide the most up-to-date information on all the Purbeck quarries.

The exact beds of stone found at each quarry vary. As the beds run north from the sea they are found nearer to the surface. At the time of the visits the two quarries in Swanage (Swanage Quarry and California Quarry) were mainly supplying grub and roach. The quarries 'on the top' north of Swanage were supplying thornback and wetsom stones predominantly.

Some of the quarries are close together and are obviously working the same deposits of stone. However, in the samples obtained for testing there was a slight difference in colour or the amount of shell in, for example, thornback from the various quarries. The actual samples from each quarry are described rather than giving one description for thornback and one for roach, etc.

All the Purbeck Limestones are shallow bedded. The depth of block is on average 150 to 200 mm but can go up to 380 mm. There is, therefore, no need to use plug and feathers to extract the stone. It can be merely lifted out using an excavator with a bucket attachment.

Reference buildings

Purbeck Marble is the most famous of the currently available hard British limestones that can be polished. It was first used, to a limited extent, by the Romans. However, it came into its own in the mid-thirteenth century when it was used extensively to form slender contrasting columns in many churches and important cathedrals. Because the beds were shallow the stone was often 'end-bedded', meaning that the natural bed of the stone was vertical in the column. This lead to failures and many old columns have been banded with metal to prevent them splitting.

Salisbury, Lincoln, Durham and Ely Cathedrals and Westminster Abbey are all decorated internally with Purbeck Marble columns. About a century later in the 1330s fashions changed and the marble was not used much until Victorian times when it was employed in restoration work. More recently Purbeck Marble was used for the Chapter House floor at Exeter Cathedral.

Purbeck Marble is not actively quarried today. Anyone wishing to use the stone should contact Mr Haysom at St Aldhelm's Quarry, who has access to a marble seam at another quarry.

Purbeck Limestone from the Middle Purbeck division has been used for every possible building purpose – paving, walling, roof tiles, etc – and can be seen all over the Isle of Purbeck. Every bed of stone is put to use and Corfe Castle bears witness to its durability.

The new town centre of Shepton Mallet, built in 1973, has random coursed Purbeck Limestone for much of the ground floor elevations and for the landscaping walls.

Purbeck Stone was used for the plinth courses at Green College Oxford in 1979.

PURBECK LIMESTONE – CALIFORNIA QUARRY

Grid reference SZ 022 776

Lessee and quarry location Mr S W Paine, California Quarry, Herston, Nr Swanage. Tel: Swanage (092-92) 3292

This large quarry is in the south-west quadrant of Swanage in the suburb of Herston. It can be reached by following the long track down towards the coast from Purbeck View. It was opened in 1965.

Petrography *(Plate 8)*
The stone from this quarry is Purbeck Limestone from the Middle Purbeck division. There are two separate faces at this quarry. At one face the beds of stone are as follows:

Rag – used for walling.

Grub – the largest blocks of stone, at this quarry, are available from this bed and are 200 to 250 mm deep. The grub is a dark blue stone with many shells.

Roach – pale grey stone.

Freestone.

Downs view.

At the other face the beds of stone rise up to be nearer the surface and the beds are thinner and more fragmented. The freestone is only 900 mm below the surface where it is shallow bedded and used for walling. In between the freestone and the downs view are the dun cow and leper beds. The latter is 730 mm deep.

The above are just a selection of some of the beds at this quarry. The samples obtained for testing were two pieces of grub, one of roach and one from the dun cow bed.

This is a typical Purbeck quarry in that all the stone is shallow bedded – usually 150 to 200 mm on bed. In between the beds of stone are fragmented, useless stones and often a great deal of clay. Not every bed within the Middle Purbeck is found at each of the quarries. At this particular quarry there are no sizeable blocks of thornback or wetsom stone.

Reference buildings
Stone from this quarry is used mainly for walling and paving. The grub is very durable in aggressive environments and was used, for example, to build sea walls at Reculver in Kent. This quarry supplies the masonry yards of Haysom's and Landers with stone.

(continued)

Results of tests   Purbeck Limestone – California Quarry

Velocity of sound (m.s^{-1})	Porosity (% volume)	Saturation coefficient	Microporosity (% saturation)	Capillarity coefficients French A (g.cm^{-2}.min$^{-1/2}$)	Belgian GC	Belgian G	Crystallisation test (% weight loss)
Grub bed							
5300	2.7	0.70	72	(1.2)	(−3.8)	(−21.3)	0, 0
5200	2.3	0.53	61	(0.08)	(0.8)	(13.0)	0, 0
Grub bed							
(4800)	1.6	0.73	57	0.09	−1.6	6.8	0, 0
5300	2.5	0.64	47	(0.1)	(−5.1)	(10.9)	0, 0
Dun cow bed							
5200	4.4	0.79	75	(0.3)	(−0.1)	(32.4)	0, 0
5300	2.1	0.56	82	(0.3)	(−0.8)	(29.1)	0, 0
Roach bed							
5100	3.6	0.61	68	−*	−*	−*	1, 0
5300	2.6	0.61	71	−*	−*	−*	0, 0

*Sample was not a suitable shape for this test

Durability
Grub bed pieces: Class A
Dun cow bed: Class A
Roach bed: Class A

PURBECK LIMESTONE – DOWNS QUARRY

Grid reference SY 973 792

Owners and quarry location Harden Bros, Downs Quarry, Kingston Road, Langton Matravers, Nr Swanage.
Tel: Worth Matravers (092-943) 255

This quarry is on the B3069 about 1 mile west of Swanage near the village of Langton Matravers.

Petrography *(Plate 8)*
This quarry works Purbeck Limestone from the Middle Purbeck division. Of the various beds of stone worked, samples were supplied for testing as follows:

Grub – pale bluish-grey stone with many pieces of small blue shell.

Thornback – grey-buff stone with blue-grey pieces of shell.

Wetsom – dense blue-grey stone with small clumps of brownish shell.

Freestone – buff-grey stone with numerous small shells which are also buff-grey in colour.

All the stone from this quarry is typically 200 to 250 mm in depth.

Reference buildings
This stone is used locally for walling.

Results of tests

Velocity of sound (m.s^{-1})	Porosity (% volume)	Saturation coefficient	Microporosity (% saturation)	French A (g.cm^{-2}.min$^{-1/2}$)	Belgian GC	Belgian G	Crystallisation test (% weight loss)
Freestone							
4400	17.1	0.49	51	(1.2)	(−4.4)	(−17.3)	2, 3
(4400)	17.5	0.55	47	1.6	−3.4	−12.9	4, 6
Thornback bed							
5300	4.1	0.59	62	(0.2)	(−3.7)	(−26.8)	0, 0
(3800)	8.8	0.48	66	0.5	−3.5	−15.4	0, 0
Wetsom bed							
(5400)	1.5	0.77	61	0.1	5.2	27.1	0, 0
5400	2.4	0.68	56	(0.2)	(−0.6)	(24.6)	0, 0
Grub bed							
5100	3.9	0.78	63	(0.3)	(1.1)	(19.9)	0, 0
(4500)	2.2	0.79	73	0.1	−1.2	32.5	0, 0

Capillarity coefficients span the French A, Belgian GC and Belgian G columns.

Durability
Freestone: Class B
Thornback bed: Class A
Wetsom bed: Class A
Grub bed: Class A

PURBECK LIMESTONE – KEATES QUARRY

Grid reference SY 982 784

Owners and quarry location K W H and H E Keates, 31 Eastington Road, Worth Matravers. Tel: Worth Matravers (092-943) 207

The quarry is on the approach road to Worth Matravers from Langton Matravers. It was opened in the 1950s and has plenty of reserves of stone.

Petrography *(Plate 8)*
The stone from this quarry is Purbeck Limestone from the Middle Purbeck division. There is 4.5 m of overburden. However, not all of this overburden is wasted; some pieces of coloured stone can be found which are suitable for crazy paving. The beds of stone are as follows:

Roach – used for walling.

Thornback – 380 mm deep bed of which the top 100 mm is mud.

Wetsom – buff-grey dense stone.

Freestone – blue-grey stone with pieces of blue shell.

Bottom bed – used for lintels.

Some large blocks of stone are available from this quarry, up to 1.8 × 1.2 m, but like all Purbecks there is no great depth on bed, 380 mm is the maximum. Samples from the wetsom and freestone beds were obtained for testing.

Reference buildings
The roach stone is used for walling for housing whilst the thornback beds are used for memorials, paving and cladding. The wetsom stone is the best for sawing and takes a good polish. The freestone is used for quoins and for carving and the bottom bed is used for lintels.

Results of tests

Velocity of sound (m.s^{-1})	Porosity (% volume)	Saturation coefficient	Microporosity (% saturation)	Capillarity coefficients French A (g.cm^{-2}.min$^{-1/2}$)	Belgian GC	Belgian G	Crystallisation test (% weight loss)
Wetsom bed							
(5400)	2.6	0.51	54	0.07	−0.5	−2.1	0, 0
(4900)	3.3	0.52	48	0.07	−2.2	−25.6	0, 0
Freestone							
(4600)	9.1	0.60	71	0.3	−0.3	4.2	0, 0
5000	8.5	0.62	71	(0.8)	(−1.3)	(−1.0)	0, 0

Durability
Wetsom bed: Class A
Freestone: Class A

PURBECK LIMESTONE – LANDERS QUARRY

Grid reference SY 978 790

Owners and quarry location Landers Quarries, Kingston Road, Langton Matravers, Swanage. Tel: Worth Matravers (092-943) 205

This quarry is on the B3069 near Langton Matravers. It was opened in 1920 and has plenty of reserves of stone. The quarry has its own masonry yard where stone from other Purbeck quarries is worked in addition to its own output.

Petrography *(Plate 8)*
The stone from this quarry is Purbeck Limestone from the Middle Purbeck division. Some of the beds found at this quarry include the freestone, wetsom and thornback. The stone is on average 300 to 380 mm on bed. Lengths of stone of up to 3 m are possible to find but the more common length is between 900 mm and 1.5 m. The average size of a block is 1200 × 900 × 300 mm deep. The samples obtained for testing were as follows:

Freestone – a grey stone with numerous small shells.

Wetsom – buff-grey stone with numerous pieces of buff coloured shell.

Thornback – blue-grey in colour with numerous pieces of buff coloured shell.

Reference buildings
All the beds provide stone suitable for most building purposes. In addition, stone from this quarry was used for the extension to Bournemouth crematorium, and for stone setts for Dover Castle, Corfe Castle and the Tower of London.

Results of tests

Velocity of sound $(m.s^{-1})$	Porosity (% volume)	Saturation coefficient	Microporosity (% saturation)	French A $(g.cm^{-2}.min^{-1/2})$	Belgian GC	Belgian G	Crystallisation test (% weight loss)
Thornback bed							
5100	3.2	0.62	54	(0.1)	(0.07)	(3.8)	0, 0
(4500)	6.3	0.47	63	0.2	0.2	6.6	0, 0
Wetsom bed							
5000	4.8	0.54	36	(0.3)	(−1.4)	(−35.0)	0, 0
(3500)	12.9	0.46	72	–	–	–	0, 0
Freestone							
5100	5.6	0.60	67	(0.2)	(−3.6)	(3.4)	0, 0
(5000)	5.4	0.56	65	0.2	−8.3	−34.6	0, 0

Capillarity coefficients span the French A, Belgian GC and Belgian G columns.

– Not tested

Durability
Thornback bed: Class A
Wetsom bed: Class A
Freestone: Class A

PURBECK LIMESTONE – SWANAGE QUARRY

Grid reference SZ 015 785

Lessee and quarry location J Suttle, Swanage Quarries, Swanage, Dorset.
Tel: Swanage (092-92) 3576

This quarry is in Swanage itself in the south-west quadrant of the town off Panorama View. The quarry is up beyond the caravan park but there is an office in Panorama Road where callers can obtain directions. This is a large quarry with several faces.

Petrography *(Plate 8)*
The stone worked at this quarry is Purbeck Limestone from the Middle Purbeck division. The beds of stone are as follows:

Royal – this is a blue stone which is about 250 mm on bed and is part of the ragstone which extends from ground level to the top of the grub bed.

Devils bed – 600 mm deep bed of a blue shelly stone.

After this bed there is muddy overburden within which is a very hard stone used for crazy paving. The next recognised bed is:

Grub – darker blue than the devils bed, the depth of this bed is 250 mm.

Roach – blue-grey shelly stone and the depth of the bed is 450 mm.

The thornback and the wetsom beds then combine to form a bed 750 mm deep.

Freestone – 1 m is the depth of this bed. The stone is blue-grey and shelly. The freestone bed is laminated and no large blocks of stone are available – 150 to 230 mm on bed is the average size.

Blocks of devils bed, grub, roach and freestone were obtained for testing.

Reference buildings
Stone from each of the beds can be put to some building use. In particular, the roach is used for cladding and for decorative work since it takes a good polish. The roach is also split for paving whilst the freestone is used for walling. Much of the stone is sent to Bournemouth and the rest goes to Surrey and Bridport. Purbeck Limestone from this quarry was used in Brentford Cathedral to match the existing Kentish Ragstone.

Results of tests

Velocity of sound (m.s^{-1})	Porosity (% volume)	Saturation coefficient	Microporosity (% saturation)	Capillarity coefficients French A (g.cm^{-2}.min$^{-1/2}$)	Belgian GC	Belgian G	Crystallisation test (% weight loss)
Roach bed							
5600	2.6	0.70	62	(0.1)	(−2.9)	(6.5)	0, 0
5500	2.8	0.79	66	(0.2)	(6.8)	(94.8)	0, 0
Devils bed							
4900	3.0	0.59	64	(0.1)	(1.9)	(29.9)	0, 1
5000	3.0	0.59	56	(0.1)	(2.8)	(39.7)	0, 0
Grub bed							
5100	1.8	0.75	68	(0.1)	(8.1)	(109.6)	0, 0
5300	2.0	0.58	62	(0.1)	(2.0)	(29.9)	0, 0
Freestone							
5000	3.6	0.61	66	(0.2)	(−1.6)	(−32.9)	0, 0
(5000)	2.6	0.61	54	0.1	−4.8	−4.1	0, 0

Durability
Roach bed: Class A
Devils bed: Class A
Grub bed: Class A
Freestone: Class A

PURBECK-PORTLAND STONE AND PURBECK MARBLE – HAYSOM'S QUARRIES

Grid reference SY 965 761 (St Aldhelm's Head)

Owners and quarry location W J Haysom and Son, St Aldhelm's Quarry, Worth Matravers, Swanage.
Tel: Worth Matravers (092-943) 217
or Swanage (092-92) 2586

St Aldhelm's Quarry is reached by following the road through the village of Worth Matravers which is off the B3069 when travelling to the west. The turn-off to the quarry is then on the left immediately before a farmyard. After following this track for half a mile or so you can see the quarry on the right in a dip in the cliffs. The quarry was opened in 1934. It has its own masonry yard where stone from other quarries is also worked.

Petrography *(Plate 8)*
The stone quarried at St Aldhelm's Head is the oolitic Purbeck-Portland Stone from the Portland Beds of Jurassic age. The stone is covered by 2.4 m of overburden. The beds of stone are then as follows:

Shrimp bed – 2.1 to 2.4 m in depth.

Blue bed – 600 mm in depth, stone used for lintels.

Spangle beds – Two beds both 1.2 m in depth. This is a hard greyish-white stone with numerous blue-grey shells.

Pond freestone – 1.5 m deep bed of a greyish-white stone with faint traces of blue-grey shell.

Samples of stone from the spangle beds and pond freestone were tested together with samples of blue and green 'marble'. The marble is not quarried at St Aldhelm's Head but at another quarry owned by the same company. It is not quarried regularly but is only worked on demand.

Reference buildings
The Purbeck-Portland Stone was used for the new Assembly Hall at Winchester School. The bowl of the font of Clifton Cathedral, Bristol, is from the spangle bed.

Results of tests

Velocity of sound (m.s^{-1})	Porosity (% volume)	Saturation coefficient	Microporosity (% saturation)	Capillarity coefficients French A (g.cm^{-2}.min$^{-1/2}$)	Belgian GC	Belgian G	Crystallisation test (% weight loss)
Spangle bed							
(4700)	8.6	0.75	82	0.7	1.9	−2.5	2, 4
4500	8.4	0.73	89	(0.7)	(0.9)	(3.0)	5, 6
Pond freestone							
(4300)	12.5	0.79	81	2.0	1.8	8.1	29, 28
4300	12.6	0.79	83	(2.5)	(1.4)	(9.1)	41, 49
Blue marble*							
–	0.9	1.00	48	(0.09)	(5.8)	(88.8)	0, 0
Green marble							
(4500)	1.6	0.75	65	0.2	17.5	233.1	0, 0
5100	1.2	0.79	52	(0.1)	(5.8)	(66.0)	0, 0

– Not tested
*Only 1 sample tested

Durability
Spangle bed: Class B
Pond freestone: Class D to E
Blue marble: Class A
Green marble: Class A

SALTERWATH STONE

Grid reference NZ 599 029

Owners Cumbria Stone Quarries Ltd, Silver Street, Crosby Ravensworth, Cumbria, CA10 3SA. Tel: Ravensworth (093-15) 227

Quarry location Nr Crosby Ravensworth, Cumbria

The Salterwath quarry is about 3 miles north-west of Orton off the B6261. Orton is not far from Junction 38 on the M6. There are old quarries nearby and also areas for future exploitation.

Petrography *(Plate 9)*
Salterwath Stone is of early Carboniferous age. It is light grey-blue in colour. Under about 3.5 m of overburden there are six beds of stone in a 180 m workable face. The average depth of stone on bed is from 300 to 410 mm.

The Salterwath and Orton Scar quarries are both worked by men from Orton.

Reference buildings
The stone will accept a good polish and can be used for various building applications of which the following are examples:

Magnum Hotels at Leicester and Manchester – flooring and internal columns, 1973
Kentmear House, Kendal – walling, 1970
Britannia Bridge, Menai Strait – cladding for bridge towers, 1979

Results of tests

Velocity of sound (m.s^{-1})	Porosity (% volume)	Saturation coefficient	Microporosity (% saturation)	Capillarity coefficients French A (g.cm^{-2}.min$^{-1/2}$)	Belgian GC	Belgian G	Crystallisation test (% weight loss)
4900	4.7	0.85	63	(0.2)	(4.3)	(60.5)	0, 0
4700	4.6	0.92	69	(0.7)	(3.1)	(15.9)	0, 0

Durability
Class A

STOWEY STONE

Grid reference ST 599 595

Owner Mr G J Bissex, 2 Market Place, Radstock, Bath. Tel: Radstock (0761) 33167

Quarry location Stowey Quarry & Lime Co Ltd, Stowey, Bishop Sutton, Nr Bristol. Tel: Radstock (0761) 52356

This quarry lies just to the east of Bishop Sutton which is on the A368 west of Bath. There are a further 3¼ hectares of ground for future quarrying.

Petrography *(Plate 9)*
Stowey Quarry has two stones, a white and a blue fine-grained limestone from the Lower Lias of early Jurassic age. The white Lias is grey rather than white and both stones have rather mottled, smooth surfaces, with no evidence of shell. There are several beds of stone but none is greater than 200 mm in depth.

Reference buildings
Stowey Stone is largely used to provide 75, 100, 125 or 180 mm courses of walling in and around Bristol. It tends to be used for larger, individual houses rather than for local authority estates. It was used at the following:

Bishopsworth Church, south-west of Bristol – new porch, 1970
Bristol Polytechnic, Clifton – new extension, 1970
Houses in Lichfield Road, Bristol – 1976
Old Clifton Down Station – garden walls, 1980
Bristol Water Works – retaining walls at Radstock (1981) and Frome (1982)

Results of tests

Velocity of sound ($m.s^{-1}$)	Porosity (% volume)	Saturation coefficient	Microporosity (% saturation)	Capillarity coefficients French A ($g.cm^{-2}.min^{-1/2}$)	Belgian GC	Belgian G	Crystallisation test (% weight loss)
White stone							
4000	3.8	0.89	59	(0.3)	(0.2)	(24.0)	0, 0
4100	4.7	0.66	68	(0.3)	(−2.6)	(2.0)	0, 0
Blue stone*							
4500	3.4	0.90	69	–⁺	–⁺	–⁺	0, 0

*Only 1 sample tested
⁺Sample was not a suitable shape for this test

Durability
White stone: Class A
Blue stone: Class A

SWALEDALE FOSSIL STONE

Grid reference NZ 218 078

Owners Cumbria Stone Quarries Ltd, Silver Street, Crosby Ravensworth, Cumbria, CA10 3SA. Tel: Ravensworth (093-15) 227

Quarry location Barton, N Yorks

The quarry is just to the east of Scotch Corner on the A1 off the minor road leading to Middleton Tyas. There are several quarries nearby which are worked for road making and lime burning. There are also further areas to exploit for building stone.

Petrography *(Plate 9)*
Swaledale Fossil Stone is of early Carboniferous age. Under 1.2 m of overburden there are seven beds of stone ranging in depth from 200 mm to 1.2 m and with varying fossil deposits. The beds are not well defined and are unpredictable. The largest bed can provide stone from between 380 and 600 mm on bed. The stone is buff-grey in colour with numerous attractive fossils.

Reference buildings
The stone is used as a decorative material and for cladding of which the following are examples:

Blackburn Cathedral – flooring, 1969
Terminal 2, Heathrow Airport – flooring of ground floor, 1973
Interland Hotel, Leeds – cladding, 1974
Manchester Cathedral – flooring, the work was done in several stages from 1975 onwards
Clydesdale Bank HQ, Glasgow – flooring, 1980

Results of tests

Velocity of sound ($m.s^{-1}$)	Porosity (% volume)	Saturation coefficient	Microporosity (% saturation)	Capillarity coefficients French A ($g.cm^{-2}.min^{-1/2}$)	Belgian GC	Belgian G	Crystallisation test (% weight loss)
(4500)	2.6	0.77	41	0.2	−0.4	0.4	0, 0
(4200)	2.7	0.77	62	0.2	0.8	15.2	0, 0

Durability
Class A

TAYNTON STONE

Grid reference SP 234 137 (Lower Farm)

Owner and quarry location Mr P R Lee, Taynton Quarries, Lower Farm, Taynton, Oxfordshire

The village of Taynton is just off the A424 between Stow-on-the-Wold and Burford about 2 miles north-west of Burford. The quarry is reached by going through the grounds of Lower Farm but prospective visitors should first write to Mr Lee. The quarry is surrounded by once famous quarries now abandoned and there are large reserves of stone.

Petrography *(Plate 9)*
Taynton Stone is an oolitic limestone from the Great Oolite of Jurassic age. It is a coarse shelly stone, buff to pale orange in colour with lighter streaks and a small bluish vein.

The stone is found under 3.0 to 3.6 m and sometimes up to 9 m of overburden. Building stone comes from the weather bed which can be up to 3 m in depth and is shelly. Within the weather bed the actual depth of stone on bed varies from 300 mm up to 750 mm. On top of the weather bed is the ragstone which is vented and unsuitable for building. In some parts of the quarry there is a 1.2 m bed which is soft, easy to work and is the very best stone. Large blocks and good lengths are available.

Reference buildings
Taynton Stone is first mentioned in the Domesday Book (1086) and was probably used one thousand years earlier by the Romans. It was largely used for building the early Oxford Colleges in the twelfth, thirteenth and fourteenth centuries, eg St John's College. In the fourteenth century 2000 tons of Taynton were used at Windsor Castle when Richard of Taynton was clerk of works. Together with Clipsham it was used for the interior of St Paul's Cathedral in the seventeenth century. In the nineteenth century Magdalen Bridge in Oxford was rebuilt in Taynton. These are just a few illustrations of Taynton's excellent record through the centuries.

In 1939 a new extension to the Bodleian Library in Broad Street, Oxford, was completed using 169 m³ of Taynton. Just before the Second World War one single block supplied 20 mm thick panelling in the hall and vestibule of Well Hall Police Station in Eltham, London. Since then the quarry has been worked single-handed. The stone is used locally for restoration work in the village, where every building is stone-walled and stone-roofed. In 1982 it was used for repairs to the ashlar of Woodstock Town Hall, Oxfordshire.

Results of tests

Velocity of sound $(m.s^{-1})$	Porosity (% volume)	Saturation coefficient	Microporosity (% saturation)	Capillarity coefficients French A $(g.cm^{-2}.min^{-1/2})$	Belgian GC	Belgian G	Crystallisation test (% weight loss)
(3000)	27.5	0.74	31	1.2	−1.3	7.6	14, 19
4100	19.5	0.67	48	(2.5)	(−1.5)	(−4.0)	10, 13

Durability
Class C

TOTTERNHOE STONE

Grid reference SP 976 224

Leased by H G Clarke & Son, 2 High Street, Weston Underwood, Nr Olney, Bucks, MK46 5JS. Tel: Bedford (0234) 711358

Quarry location and owners The Totternhoe Lime and Stone Co Ltd, Totternhoe, Nr Dunstable, Beds

The quarry is on the northern edge of the village of Totternhoe, which is just off the B489, 1 mile west of Dunstable. It is reached by taking the turning marked 'Lime Kilns only'.

The owners of this vast quarry crush most of the output or else burn it in kilns.

The building stone is extracted from one small part of the quarry. Until Mr Clarke took on the lease, in October 1981, Totternhoe building stone was only worked intermittently every few years when stone was required by Woburn Abbey. The quarry was reopened too late to be included in the testing programme but Woburn Abbey kindly donated some stone for testing.

Petrography *(Plate 9)*
Totternhoe Stone is a chalk from the Lower Chalk of Cretaceous age. It is a greyish-white colour often with a greenish tinge. The latter is due to the presence of glauconite, the potassium and iron aluminium silicate mineral also found in Kentish Ragstone. The stone has a gritty texture due to the presence of small shell fragments.

The building stone lies under 9 to 15 m of overburden and was in fact mined in the past. However, the overburden has all been removed for crushing and the stone is pulled out of the adits (entrances to the mines) by crane. There is no need to use plug and feathers to extract it. The face of building stone is between 4.2 and 5.4 m deep. The stone is very well jointed and there are no recognised beds. Some very large boulders can be obtained — up to 20 tons, but 5 tons is more common. The stone can be up to 1.8 m deep on bed.

The stone towards the top of the face contains numerous hard pyrites. These are small nodules of iron sulphide which, if left in the stone, break down to form sulphuric acid which attacks the stone. The stone towards the bottom of the face contains few pyrites and is altogether a harder and more durable stone.

Building stone from the Lower Chalk is generally known as Clunch. Strictly, in geological terms, this term only applies to the stone that came from Cambridgeshire. However, Totternhoe is now the only source of chalk used as a building stone.

Reference buildings
Totternhoe Stone was used extensively in and around its own locality in the Middle Ages as it was the only stone available. It was used to build Woburn Abbey in Bedfordshire and Ashridge House in Hertfordshire. Today Woburn Abbey uses most of the output of the quarry for its large and continuous restoration programme. Experience at Woburn and elsewhere has shown that the stone is best suited for use as ashlar, although it has been used extensively for moulded work. Its performance relates very much to the way it has been extracted, seasoned, and laid in the building. Some stone acquires a remarkable toughness after weathering.

Plate 1 Green College Oxford built in 1979 using several British limestones (see text pp 13, 25, 26, 60, 78)

Plate 2 A typical quarry face showing well jointed and bedded stone. The vertical marks are the holes left after extracting blocks with plug and feathers

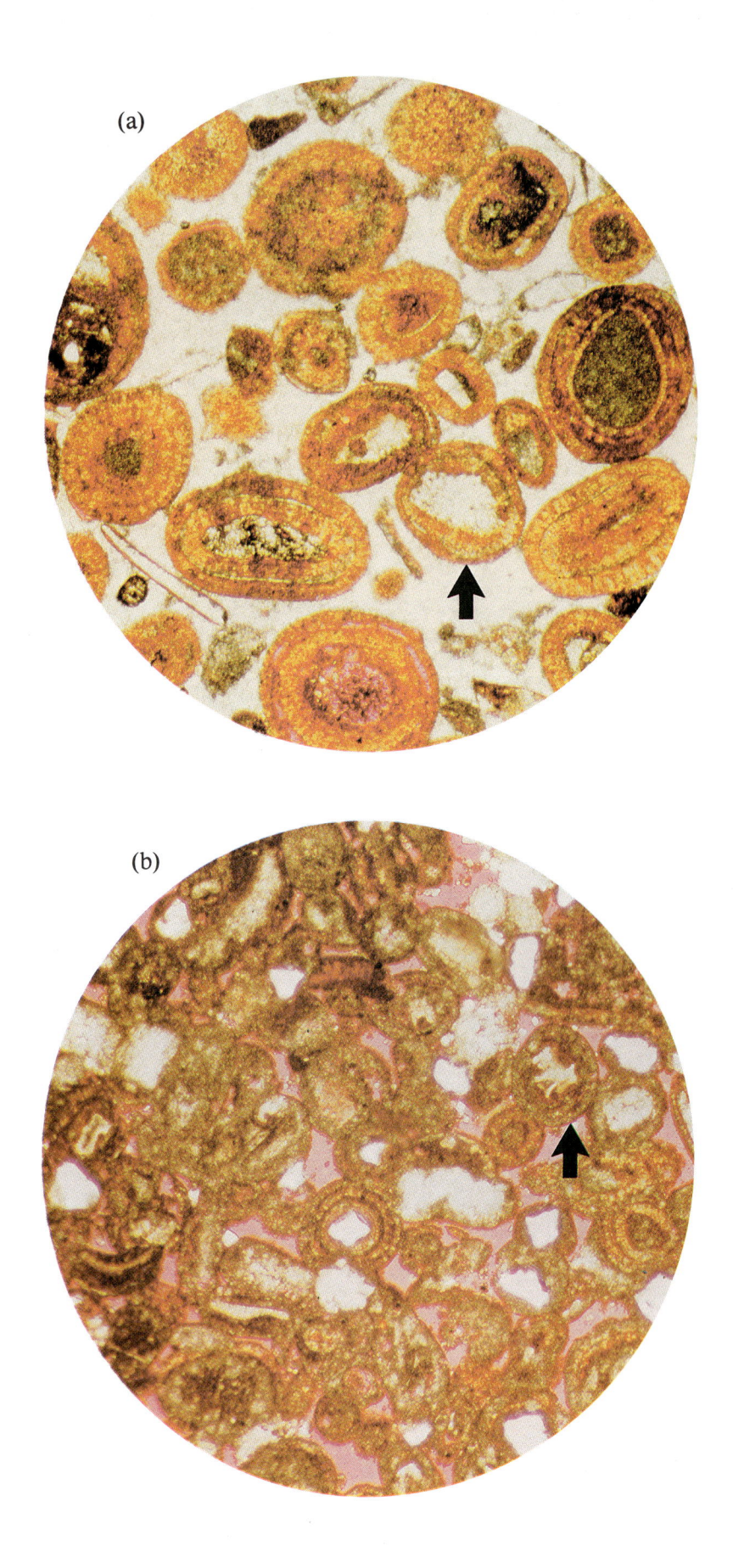

Plate 3 Photomicrographs of petrological thin sections of oolitic limestones with their pores filled with pink resin. The white areas are the solid calcium carbonate matrix; arrows indicate an oolith.
(Magnification ×45)
(a) Except for the oolith at the bottom of the picture in which the pores can easily be seen, the pores in the other ooliths are so small as to be indistinguishable
(b) The large pink pores are easily distinguishable from the ooliths and solid calcium carbonate

Ancaster weather bed

Ancaster hard white

Ancaster freestone

Ballinasloe

Bath – Combe Down – Upper Lawns Quarry

Bath – Monks Park

Bath – Stoke Ground Base Bed

Bath – Westwood Ground

Clipsham – Big Pits Quarry

Clipsham – Medwells Quarry: creamy-brown stone

Clipsham – Medwells Quarry: blue stone

Daglingworth

Plate 4 Some British limestones (actual size)

Doulting – top bed

Fenacre

Guiting – Coscombe Quarry: white walling

Guiting – Coscombe Quarry: yellow-orangey stone

Guiting – Cotswold Hill Quarry: cream-yellow stone

Ham Hill

Happylands

Hopton Wood

Hornton – blue rag stone

Hornton – brown stone

Hovingham – white stone

Hovingham – blue stone

Plate 5 Some British limestones (actual size)

Kentish Ragstone

Ketton – 'new' stone

Kilkenny Marble

Lecarrow

Mandale

Nash Rocks – Strinds Quarry

Navan

Orton Scar

Penmon (black)

Plate 6 Some British limestones (actual size)

Portland – Bowyers Quarry: whit bed

Portland – Coombefield Quarry: roach bed

Portland – Fancy Beach Quarry: roach bed

Portland – Fancy Beach Quarry: whit bed

Portland – Weston Quarry: whit bed

Portland – Weston Quarry: base bed

Portland – Sheat Quarry: roach bed

Plate 7 Portland stones (actual size)

Purbeck Limestone – California
Quarry: roach bed

Purbeck Limestone – Downs
Quarry: thornback bed

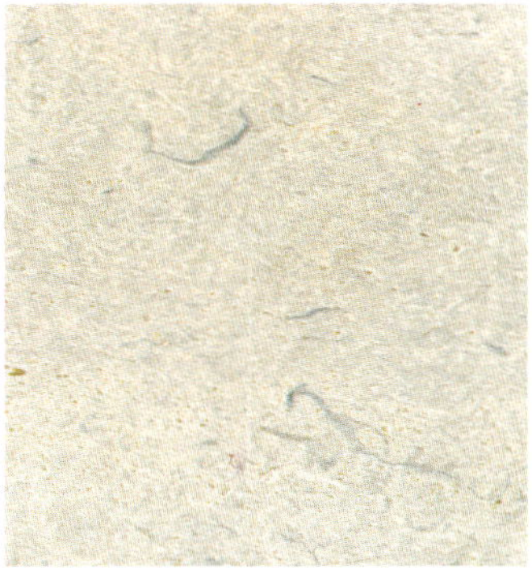

Purbeck Limestone – Keates
Quarry: wetsom bed

Purbeck Limestone – Landers
Quarry: freestone

Purbeck Limestone – Swanage
Quarry: grub bed

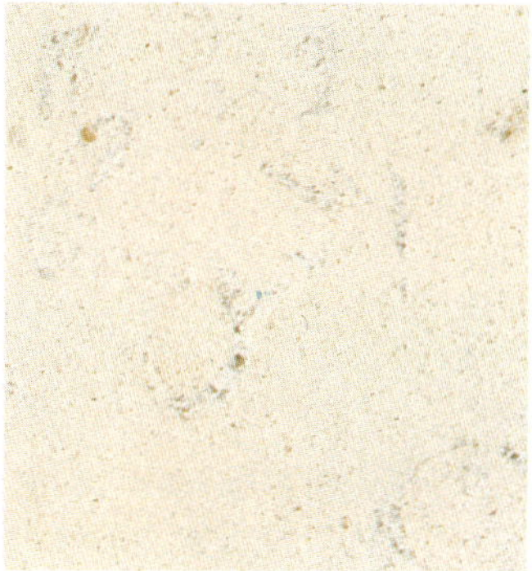

Purbeck-Portland – Haysom's
Quarry: spangle bed

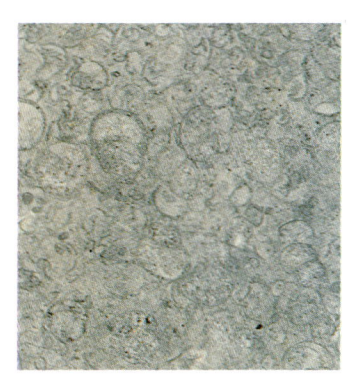

Purbeck Marble (blue)

Plate 8 Purbeck stones (actual size)

Salterwath

Stowey – white stone

Stowey – blue stone

Swaledale Fossil

Taynton

Totternhoe

Ulverston Marble – oatmeal stone

Ulverston Marble – mottled stone

Weldon – fine-grained stone

Plate 9 Some British limestones (actual size)

Totternhoe was used extensively in the local churches of which the west front of Dunstable Priory is an example although, unfortunately, this is somewhat decayed. Another example is in the north arcade of the church at Eaton Bray, the next village to Totternhoe. It was also used for the reredos at St Albans Cathedral and the organ screen at Peterborough Cathedral, where indoors of course it survives well. Totternhoe was used for the Roman medallions at Salisbury Hall, just south of London Colney on the A6 in Hertfordshire, and at Audley End House near Saffron Waldon, Essex.

In 1982 the stone was used for restoration at Hitchin Priory for ashlar over the back entrance. Also in 1982 it was used for a new extension on the river side of the Swan Hotel, Bedford.

Results of tests

Velocity of sound ($m.s^{-1}$)	Porosity (% volume)	Saturation coefficient	Microporosity (% saturation)	Capillarity coefficients French A ($g.cm^{-2}.min^{-1/2}$)	Belgian GC	Belgian G	Crystallisation test (% weight loss)
2800	25.7	0.87	89	(8.4)	(3.8)	(25.8)	Failed after
(2600)	27.3	0.90	85	–	–	–	8 cycles

– Not tested

Durability
Class F

Although we have not been able to test the current output of the quarry, it is unlikely to differ from Class F.

ULVERSTON MARBLE

Grid reference SD 288 723 (Baycliff)

Owners Burlington Slate Ltd, Cavendish House, Coniston, Cumbria, LA21 8ET. Tel: Coniston (096-64) 515

Quarry location Skiddaw Quarry, Baycliff, South of Ulverston, Cumbria

Skiddaw Quarry is at Baycliff which is just off the A5087. It is an old aggregate quarry which was last worked in the 1950s. It was reopened in November 1981 when, for the first time, the stone was extracted for building purposes rather than for crushing. There are large reserves of stone.

Before November 1981 the company extracted stone from a quarry at nearby Stainton-with-Adgarley. This stone was also known as Ulverston Marble. However, this quarry is no longer worked and all Ulverston Marble now comes from Skiddaw Quarry. This quarry was not visited although samples were provided for testing.

Petrography *(Plate 9)*
Ulverston Marble is of Carboniferous age. It is not a marble in the geological sense, the term here being used to indicate a hard limestone which will take a polish.

The total depth of the quarry face is 12 m. The first 3 to 4.5 m is overburden from which a walling stone can be extracted. There are then several 1.2 m beds of the two building stones laid down over one another. One stone is an oatmeal or dark cream colour and the other is a light brown mottled stone. There is very little shell evident in either stone. Both stones can be obtained at depths of 830 mm on bed.

This quarry was opened too late to be included in the full testing programme nor are there yet any reference buildings where the stone can be seen.

Results of tests

Porosity (% volume)	Saturation coefficient	Crystallisation test (% weight loss)
Oatmeal stone		
3.7	0.65	0, 1
		0, 0
Mottled stone		
1.2	0.88	0, 0
		0, 0

Durability
Oatmeal stone: Class A
Mottled stone: Class A

WELDON STONE

Grid reference SP 919 888

Owners Weldon Stone Enterprises Ltd, The Menagerie, Park Hill, Castle Ashby, Northampton. Tel: Yardley Hastings (060-129) 252

Quarry location Weldon, Nr Corby. Tel: Corby (053-63) 61545

The quarry is on the north-west side of the A43 about 0.5 mile south of the village of Weldon and 2 miles east of Corby. This stone has been quarried for centuries from various workings. There is a large area of 12 hectares for future work. The present quarry was opened in 1977.

Petrography *(Plate 9)*
Weldon Stone is an oolitic limestone from the Lincolnshire Limestone formation of middle Jurassic age. It is brownish buff in colour and there is a fine-grained stone as well as a more shelly coarse variety. The freestone is found under 1.8 m of overburden within which there is a ragstone which is used for walling. The maximum depth of freestone on bed is 900 to 1200 mm and large blocks can be supplied.

There are no well defined beds of stone around the quarry face. The freestone blocks occur haphazardly. It is thought that below the present floor of the quarry there will be a blue shelly rag, then a further layer of freestone and finally Weldon 'marble' — a blue shelly limestone.

Reference buildings
Weldon Stone together with Ancaster, Clipsham and Ketton stones, are the best of the Lincolnshire Limestones quarried today. Barnack, King's Cliffe and Stamford are some of this famous and durable group of limestones that are, unfortunately, no longer available. These stones provided character for the area in and around Stamford just as Cotswold stone had for its area. One of the earliest examples of Weldon Stone is the nearby Queen Eleanor Cross at Geddington, south-west of Corby, erected in 1295. The Lincolnshire Limestones have also been very successful further afield. Weldon was used to build the old St Paul's Cathedral before the Great Fire.

Weldon was first introduced to Cambridge in the fifteenth century when it was used at King's College Chapel together with Clipsham and a magnesian limestone. Work began at the east end where a white magnesian limestone from Yorkshire was used. However, building was interrupted by war and the west part was later finished in Weldon. It is possible to follow the course of the contract as the darker stone replaces the whiter, northern stone. The magnificent vaulting in the Chapel is in Weldon Stone.

It was used extensively to build Jesus College Cambridge on the site of the nunnery of St Radegond. This was given to Bishop Alcock by Henry VIII after the behaviour of the inhabitants had become far from nun-like. Together with other stones Weldon was used for the rebuilding of Great St Mary's Church, for Gonville and Caius College and for dressings at Trinity College. These last three examples are from the sixteenth century. In the early twentieth century Weldon was used for extensions at Sidney Sussex and Gonville and Caius Colleges. Weldon cladding was used for the new administrative building for Wolfson College in Barton Road completed in the late 1970s. This building has a Clipsham Stone plinth.

The dressings of the brick towers of the Bishop's Palace at Ely, which was also built by Alcock, are of Weldon.

(continued)

In Oxford, Weldon was used, in the early twentieth century, for ashlar in St Alban's Quadrangle and the Warden's House at Merton and for the Rhodes Building at Oriel. A large Oxford contract in 1979 involved the use of Weldon ashlar at the new Green College. On the same site, the Old Observatory was refaced in Weldon in the 1960s. Also in the 1960s the stone was used for restoration work at Norwich and Ely Cathedrals.

Weldon was used for restoration at Rochester Cathedral and the Chapter House, Lincoln.

Another large contract, in 1981, involving Weldon can be seen in the Market Square, Northampton. The new building for C & A Modes which fronts the Welsh House and serves as an entrance to the Grosvenor Centre shopping complex is clad in Weldon. This matches the Weldon Stone surrounds to the windows of the Welsh House itself which has Hornton Stone ashlar and was restored in 1972.

Weldon was used in 1981 for the tracery windows at Bedford School.

Results of tests   Weldon Stone

Velocity of sound ($m.s^{-1}$)	Porosity (% volume)	Saturation coefficient	Microporosity (% saturation)	Capillarity coefficients French A ($g.cm^{-2}.min^{-1/2}$)	Belgian GC	Belgian G	Crystallisation test (% weight loss)
Fine-grained stone							
2300	27.4	0.68	30	(7.2)	(−0.7)	(0.9)	9, 8
(2900)	26.8	0.67	27	17.1)	−2.0	4.0	13, 11
Coarse-grained stone							
1800	27.7	0.65	22	(∞)	(−0.2)	(−2.6)	12, 9
2000	26.6	0.66	32	(3.9)	(−0.7)	(−0.3)	7, 8

Durability
Fine-grained stone: Class C
Coarse-grained stone: Class C

WROXTON STONE

Grid reference SP 385 428

Leased by Mr P Miers, Marine Cottage, Horley, Nr Banbury, Oxon.
Tel: Wroxton St Mary (029-573) 469

Quarry location Alkerton, Nr Banbury, Oxon

The quarry is at Alkerton which is north-west of Banbury. The entrance to the quarry is 90 metres off the A422 by an Oxfordshire County Council rubbish tip. The quarry has been worked since 1918 mainly for iron ore and roadstone. There are plenty of reserves. This quarry was not visited nor were any samples tested.

Petrography

Wroxton Stone is from the Middle Lias division of Jurassic age. It is greenish-blue or brown or can be a combination of all of these colours. The depth of the face is 10 m. The average depth of stone on bed is 600 mm although blocks of nearly 1 m are available.

Reference buildings

Wroxton Stone is mainly used locally as random walling for housing. In 1982 it was used for four new houses on the right-hand side of the road when you enter the village of Hornton from the south. In the same year it was used for restoration and for piecing-in at the Old Bake House, Shutford.

Appendix A Description of the tests

This Appendix describes the procedures for each of the tests which were carried out.

VELOCITY OF SOUND

The aim of the velocity of sound test is to measure the time it takes for a wave of sound to travel down a known length of stone. Figure A1 is a graph of porosity against the velocity of sound measurement for 19 limestones. It shows that, broadly, as the porosity decreases, the velocity of sound increases. However, the correlation is not good enough to permit the estimation of one parameter from the other.

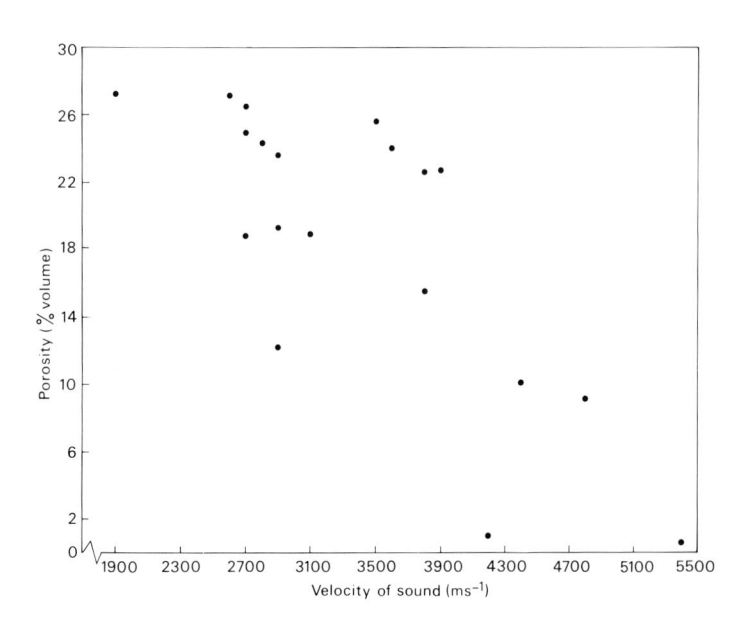

Figure A1 Graph of porosity plotted against velocity of sound for 19 limestones

The apparatus used consists of two probes connected by cables to a central unit. The probes are placed at opposite ends of the stone specimen; one acts as a transmitter, the other as a receiver. The time taken for a sound wave to travel between the two probes is displayed electronically on the central unit. Typical equipment is illustrated in reference 2.

POROSITY

Porosity is defined as the volume of a stone's pore space expressed as a percentage of the stone's total volume. It is generally measured by saturating samples with water under vacuum[3].

The samples are dried and weighed (W_0) and then saturated with water under vacuum. They are then weighed in water (W_1) and again in air (W_2) and the porosity, P, is calculated as follows:

$$P = \frac{W_2 - W_0}{W_2 - W_1} \times 100\ (\%)$$

SATURATION COEFFICIENT

The saturation coefficient is a measure of the extent to which the pores become filled when a stone is allowed to absorb water for a standard time. Once the porosity has been measured the sample is dried, immersed in water for 24 hours and weighed (W_3). The saturation coefficient, S, is then calculated as follows:

$$S = \frac{W_3 - W_0}{W_2 - W_0}$$

MICROPOROSITY

Microporosity is defined as the volume of water retained (expressed as a percentage of the total available pore space) when a suction equivalent to a 6.4 m head of water is applied to the specimen. In essence it measures the percentage of pores with an effective diameter of less than 5 μm. These very small pores are able to retain water against the applied suction and these are the pores that determine resistance to damage by frost or damage by the crystallisation of soluble salts. The higher the microporosity, the more small pores the stone contains. It is measured as follows.

Small discs of stone are dried and weighed (W_0), then vacuum-saturated with water and weighed again (W_1). The discs are placed on a porous plate which is subjected to a negative pressure equivalent to 6.4 m of water. They are weighed daily until constant weight (W_2) is reached[8]. The microporosity, M, is then calculated as follows:

$$M = \frac{W_2 - W_0}{W_1 - W_0} \times 100\ (\%)$$

CAPILLARITY

The capillarity test measures the rate at which a sample soaks up water. The test consists of standing prisms of stone of uniform cross-section, which are 100 mm in height, in a shallow tray of water. The uptake of water is monitored by following the increase in weight of the sample. The test is the subject of Belgian Standard NBN B 05-201[9], which gives full details of the experimental procedures; French Standard NF B 10-502[10] describes a very similar procedure. The test carried out at BRS was modified in two respects. Firstly, the samples were only immersed to a depth of 2 mm to reduce water uptake through the sides of the samples. Secondly, the weighings were timed accurately to within half a minute for the first few readings.

In the test, porous stones characteristically show a two-stage uptake of water. Initially, water rises up the most accessible pores to the top of the samples which show a sharp weight increase against time — the gradient α_1 of Figure A2. The weight increase is expressed as the water content, S:

$$S = \frac{W_t - W_0}{W_s - W_0} \times 100 \, (\%)$$

where W_t = weight at time t
W_s = vacuum saturated weight
W_0 = dry weight.

In the second stage of water uptake (gradient α_2 of Figure A2) water begins to fill the remaining pore space. The water content at the point of inflection is denoted by S_t.

In low-porosity stones water uptake is a much slower process, and during the time of the test water does not reach the top of the sample (see Figure A3). Here S_t is calculated at $\sqrt{t} = 100$ min$^{1/2}$, an arbitrary figure specified in the Belgian Standard.

A computer program was used to calculate the results. The computer gives the 'best-fit' line for each set of readings, plotting out S against $\sqrt{t}$ for every sample. The computer operator then decides whether the stone has shown a two-stage uptake of water as in Figure A2 or whether uptake follows the pattern of Figure A3. In the former case the computer is told which data points constitute the first slope and which the second slope and it can then work out S_t, α_1 and α_2. In the latter case it works out S_t and α_1 at the point where $\sqrt{t} = 100$ min$^{1/2}$.

However, there are some drawbacks to this test. For example, one sample of a particular stone gave a two-slope graph as in Figure A2 whilst another sample cut from the same block gave only one slope — as in Figure A3. This is fully discussed by Leary[5]. Further it was sometimes difficult to determine where the point of inflection actually occurred.

Description of Belgian and French capillarity coefficients

Apart from the coefficient G defined in Belgian Standard NBN B 05-201[9] the Belgians determine a second coefficient[11], GC.

Belgian capillarity coefficient, G
G is defined as:
  $S_t + 178\alpha_2 - 81$
and for low-porosity stones, ie one-slope graphs:
  $G = S_t + 178\alpha_1 - 81$
calculated at $\sqrt{t} = 100$ min$^{1/2}$.

If G is less than zero then the stone is considered to be resistant to frost.

Belgian capillarity coefficient, GC
GC is defined as:
  $-14.53 - 0.31\alpha_1 + 0.203S_t$
and for low-porosity stones, ie one-slope graphs:
  $GC = -6.35 + 21.47\alpha_1$

Using GC, five classes of exposure are defined:

$GC < -2.5$	Class D — stone survives four winters in a tray of sand or at least two winters at Bertrix (an inland site in the south-east of Belgium)
$GC > -2.5$ to -0.95	Class C — stone survives four winters as a paving stone

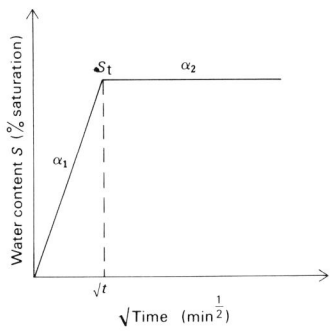

Figure A2

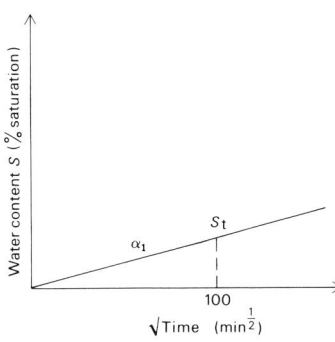

Figure A3

$GC > -0.95$ to 0 Class B — stone survives free standing for four winters

$GC > 0$ to 4.5 Class A — stone survives in a vertical wall for four winters

$GC > 4.5$ Class O — stone does not survive in a vertical wall.

French capillarity coefficient, A

A is defined as:
$$100 \, m/\sqrt{t}$$
where m = mass of water absorbed per unit area
t = time.

The French use A, together with the porosity and saturation coefficient or a freeze/thaw test, to define four zones of use for a stone[2]. A wear test is also included if the stone is required for paving, steps or thresholds. These zones are illustrated in Figure 2, Chapter 2.

One unsatisfactory feature of the French Standard (NF B 10-502) for this test is that it does not specify the range over which the initial slope of the graph (see Figure A2) should be measured. Since the initial portion is not always linear, this calls for a somewhat subjective assessment.

It should be noted that the size of the samples used was not the same as those specified in the Standard. However, this should not affect the results.

CRYSTALLISATION TEST

The crystallisation test measures the ability of a stone to withstand the pressures created by the repeated crystallisation and hydration of soluble salts within its pores.

This is achieved by subjecting samples to 15 cycles of immersion in a 14 per cent solution of sodium sulphate decahydrate and then drying in a humid oven. The weight loss is determined and compared with the losses from stones of known durability which are included in the same test.

The results are very sensitive to slight changes in experimental procedure. Full details of the recommended procedure are given in *BRE Digest* 269[12].

Appendix B Appraisal of European test procedures

This Appendix discusses two European tests – the measurement of the velocity of sound and the capillarity test – and assesses whether they have any contribution to make to the durability classification of British limestones. The results from these tests are compared with the results from the crystallisation test which has traditionally been regarded as the test which best reflects the durability of limestones.

VELOCITY OF SOUND
An early French Standard, NF B 10-001[13], divided limestones into a number of hardness categories depending on their compressive strengths and apparent densities. The categories were allocated numbers from 1 to 14, where 14 represented a very hard stone, and these were referred to as the AFNOR hardness numbers (AFNOR = Association Française de Normalisation). This classification was sometimes taken as an indication of durability but there is in fact no correlation between the AFNOR number and durability. Nowadays, a combination of the velocity of sound measurement and the surface hardness replaces the compressive strength in the assessment of the AFNOR number. Further, the AFNOR number is only used now to indicate the workability of the stone from the mason's point of view[2].

Thus, the velocity of sound measurement, on its own, gives no indication of the durability of British limestones. Indeed, Figure B1*, which is a graph of the velocity of sound plotted against the crystallisation losses for a number of stones, shows no correlation between the two parameters.

CAPILLARITY TEST
Belgian capillarity coefficient, G
$G = S_t + 178\alpha_2 - 81$ or, for low-porosity stones,
$G = S_t + 178\alpha_1 - 81$

There are two variables in G: S_t and α_2 (α_1 for low-porosity stones) (see Figures A2 and A3 in Appendix A). The quantity S_t represents, in practice, the percentage of pore space that can be filled during the initial absorption of water. One might reasonably expect some correlation between S_t and the saturation coefficient and this is borne out in Figure B2.

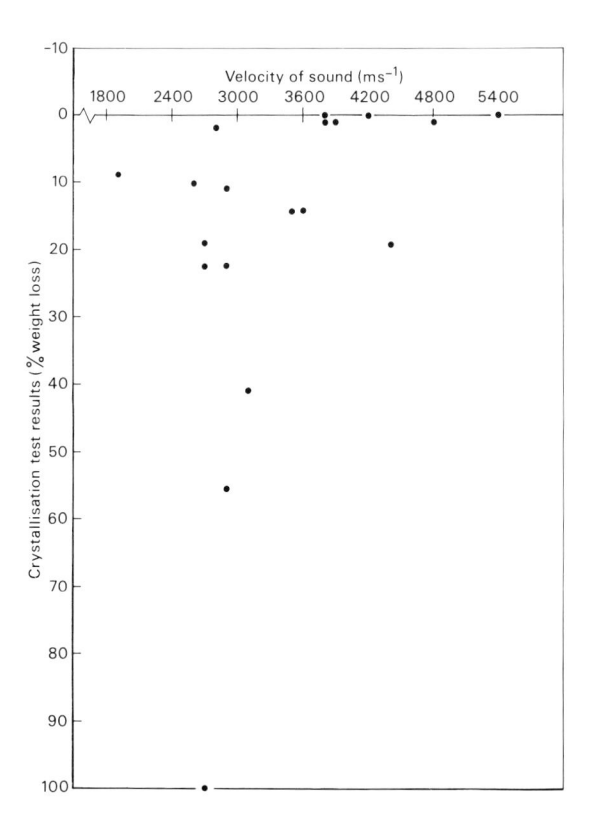

Figure B1 Graph of crystallisation test results plotted against velocity of sound

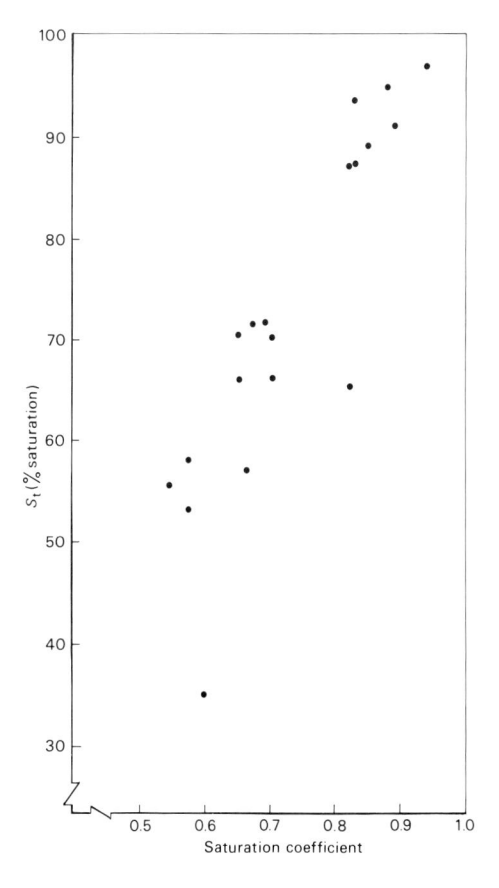

Figure B2 Graph of S_t plotted against saturation coefficient

---
*The same representative sample from the stones tested in the survey has been used to illustrate all the figures in Appendix B – see reference 5 for further details.

The variable α_2 indicates the rate at which water is absorbed into those pores which were 'by-passed' during the initial absorption. A high value of α_2, and consequently a high value of G, would indicate that the stone would ultimately reach high levels of saturation during natural exposure and thus be susceptible to frost.

Low-porosity stones contain few pores so uptake of water is slow; thus α_1 and G are both low. Such stones would be unlikely to reach high levels of saturation during natural exposure.

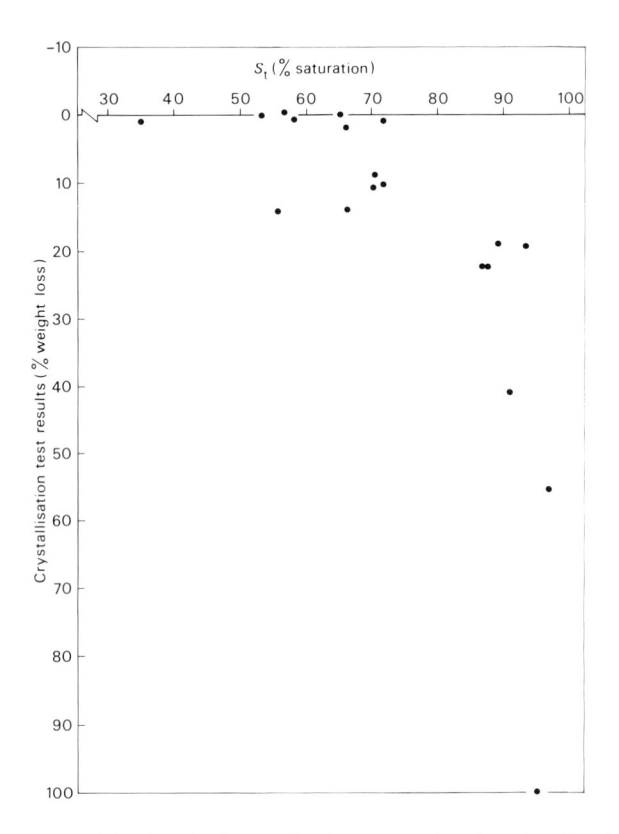

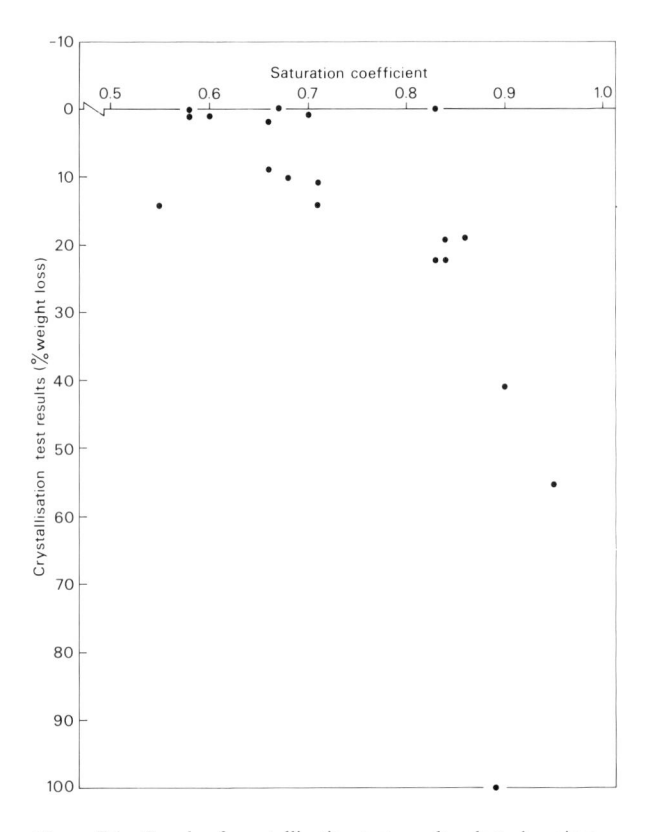

Figures B3, B4 and B5 show the crystallisation test results plotted against S_r, saturation coefficient and G respectively. The three figures are broadly the same and show that the added sophistication of S_r and G provides no better correlation than the simpler saturation coefficient.

Belgian capillarity coefficient, GC

$GC = -14.53 - 0.31\alpha_1 + 0.203S_r$ or, for low-porosity stones, $GC = -6.35 + 21.47\alpha_1$

Figure B6 plots GC against the crystallisation test results and also indicates the five classes of exposure D, C, B, A and O defined in Appendix A pp 81–82. The figure is not substantially different from Figure B4, which plots the crystallisation test results against saturation coefficient. This is not surprising since the contribution made to GC by the term $0.31\alpha_1$ is small. Further, saturation coefficient correlates closely with S_r. Unlike the coefficient G, it is difficult to see any theoretical rationale for the definition of GC. But like G, GC appears to offer no improvement over the much simpler saturation coefficient.

French capillarity coefficient, A

$A = 100 \, m/\sqrt{t}$

The French do not use A as an indicator of durability. It is used to ensure that a particularly absorptive stone is not used in harsh exposures where there are high levels of saturation. The coefficient is used, together with several other parameters, to determine the suitability of stone for use in one of four building zones (the zones are illustrated in Figure 2, Chapter 2).

In the survey, the coefficient A seldom affected the zone into which the stones fell, the influence of the other parameters predominating. Where it did influence the zone it tended to downgrade the use of stone which the crystallisation test had shown to be durable. Figure B7 plots the crystallisation test results against the zones allocated by the French system. It shows that the French system broadly placed the stones in accordance with the crystallisation test results. However, the figure also shows the downgrading of some of the most durable stones and the fact that the system lacks the sensitivity to differentiate between the middle durability stones.

CONCLUSION

Neither the velocity of sound measurement nor the capillarity test, with its three coefficients, offers any improvement over existing procedures for assessing the durability of British limestones.

Figure B3 Graph of crystallisation test results plotted against S_r

Figure B4 Graph of crystallisation test results plotted against saturation coefficient

Figure B5 Graph of crystallisation test results plotted against Belgian *G*

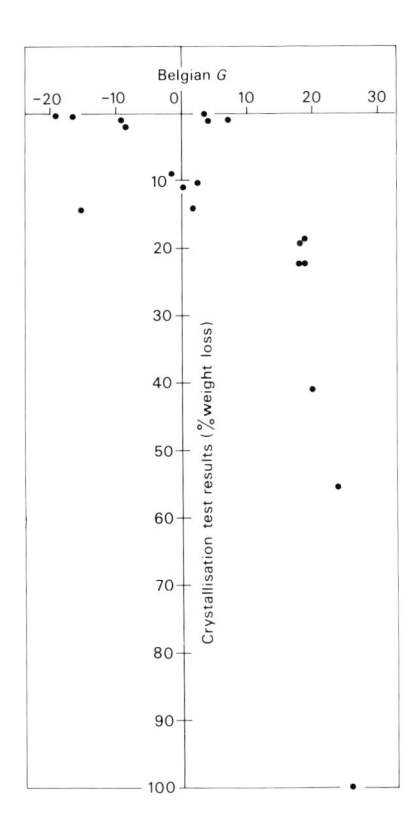

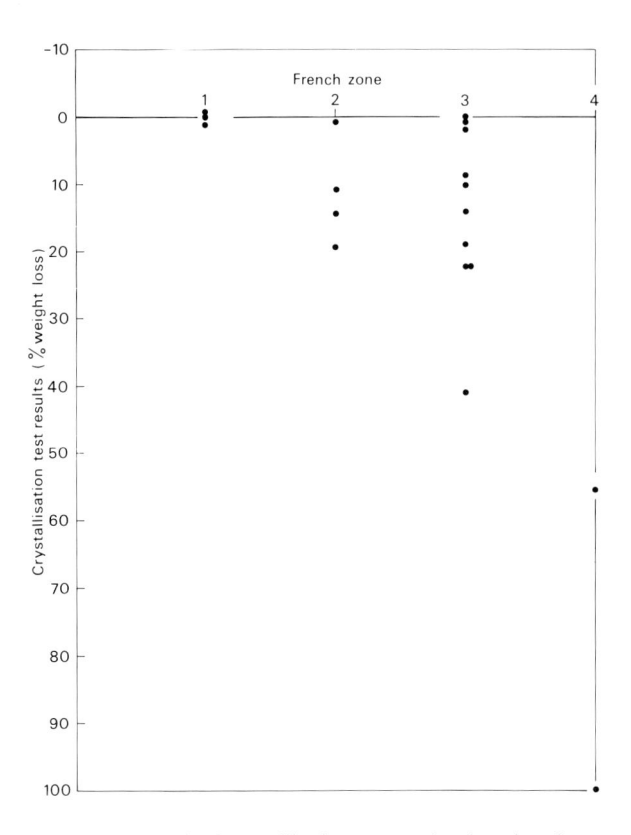

Figure B7 Graph of crystallisation test results plotted against French zones of a building

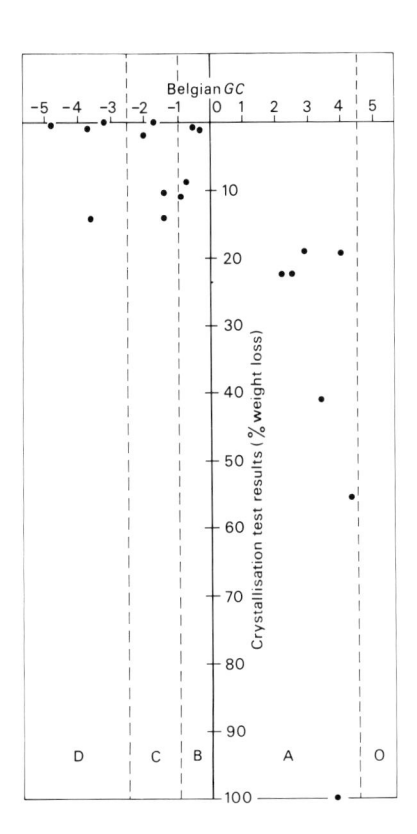

Figure B6 Graph of crystallisation test results plotted against Belgian *GC*. D, C, B, A and O are the five classes of exposure

Acknowledgements

I am indebted to several people for help and advice given during the survey: in particular I am very grateful to Martyn Owen of the Geological Museum for his advice, and to Bryan Hodgeman of the Directorate of Ancient Monuments and Historic Buildings who visited some of the quarries. David Honeyborne, John Ashurst of DAMHB, and Brian Clarke of the Stone Federation provided valuable comments at the draft stage of the book. I am further indebted to John Ashurst for providing the original ideas for Figures 1 and 2.

I wish to acknowledge Alec Clifton-Taylor for permission to quote many of the reference buildings from his excellent book *The pattern of English building*. I also acknowledge permission given by the Headmaster of Prior Park College and the Bursar of Green College Oxford to photograph the buildings.

Finally and not least, my thanks go to colleagues Clifford Price and Keith Ross.

References

1. *Natural stone directory.* Published by Ealing Publications Ltd, 73a High Street, Maidenhead, Berkshire, SL6 1JX.

2. **Honeyborne D B.** *The building limestones of France.* Building Research Establishment Report. London, HMSO, 1982.

3. **Price C A.** Testing porous building stone. *The Architects' Journal*, 1975, **162** (33) 337–339.

4. **Mamillan M.** Nouvelles connaissances pour l'utilisation et la protection des pierres de construction. *Annales de l'Institut Technique du Bâtiment et des Travaux Publics, Serie Matériaux,* No 48, Supplément No 335, Janvier 1976, pp 18–48. (Published by S A Le Batiment, 6 rue Paul-Valéry, 75116 Paris.)

5. **Leary E.** A preliminary assessment of capillarity tests as indicators of the durability of British limestones. *The conservation of stone, II.* Preprints of the Contributions to the International Symposium, Bologna, 27–30 October 1981. Part A, Deterioration. Published by Centro per la Conservazione delle Sculture all'Aperto, Bologna. pp 73–90.

6. **Honeyborne D B and Harris P B.** The structure of porous building stone and its relation to weathering behaviour. *Proceedings Tenth Symposium of the Colston Research Society,* Eds D H Everett and F S Stone. London, Butterworths Scientific Publications, 1958. pp 343–365.

7. **Cooke R U and Doornkamp J C.** *Geomorphology in environmental management.* Oxford, Clarendon Press, 1974. pp 281–283.

8. **Croney D, Coleman J D and Bridge P M.** The suction of moisture held in soil and other porous materials. *Road Research Technical Paper* 24. London, HMSO, 1952.

9. *Norme Belge* NBN B 05-201. Gélivité capacité d'impregnation d'eau par capillarité. (Resistance of materials to freezing – Water absorption by capillarity.) Brussels, Institut Belge de Normalisation, 1976.

10. *Norme Française* NF B 10-502. Produits de carrieres – Pierres calcaires – Mesure de l'absorption d'eau par capillarité. (Quarry products – Limestones – Measurement of water absorption by capillarity.) Paris, Association Française de Normalisation (AFNOR), 1973.

11. **Longuet M.** Mise au point des methodes d'appreciation de la resistance au gel des matériaux poreux. *Silicates Industriels, Mons, Belgium,* 1980, **7-8**.

12. **Building Research Establishment.** The selection of natural building stone. *BRE Digest* 269. London, HMSO, 1983.

13. *Norme Française* NF B 10-001. Matériaux – Pierres calcaires. Paris, Association Française de Normalisation (AFNOR), 1945.

Bibliography

Arkell W J. *Oxford stone.* London, Faber and Faber, 1946.

Ashurst J and Dimes F G. *Stone in building: its use and potential today.* London, Architectural Press, 1977.

Clifton-Taylor A. *The pattern of English building.* London, Faber and Faber, 1972.

Elsden J V and Howe J A. *The stones of London.* London, Colliery Guardian Co, 1923.

Hadfield J (Ed). *The new Shell guide to England.* London, Rainbird Publishing Co, 1982 (2nd edn).

Howe J A. *The geology of building stones.* London, Edward Arnold, 1910.

Oakeshott W F (Ed). *Oxford stone restored.* Trustees of the Oxford Historic Buildings Fund, 1975.

Purcell D. *Cambridge stone.* London, Faber and Faber, 1967.

Schaffer R J. *The weathering of natural building stones.* Building Research Special Report No 18. Garston, Building Research Establishment, 1972 (Facsimile reprint).

Shore B C G. *Stones of Britain.* London, Leonard Hill (Books) Ltd, 1957.

Smith E and Cook O. *Prospect of Cambridge.* London, B T Batsford Ltd, 1965.

Watson J. *British and foreign building stones.* Cambridge University Press, 1911.

Index of places where limestones may be seen in buildings

INTRODUCTORY NOTES
The index covers the reference buildings named in Chapter 3. The stone present in the building is named, not its quarry.

Abingdon, Cosenor's House: Guiting, 30
Ampleforth Abbey Church: Hornton, 37
Anglesey: Penmon, 49
    Dovecot nr Penmon quarry: Penmon, 49
Armagh: Navan, 47
Ashbourne
    Church: Lecarrow, 44
    'Peveril-of-the-Peak' Hotel: Hovingham, 38
Ashford in the Water, nr Bakewell: Mandale, 45
Ashridge House: Totternhoe, 74
Audley End House, nr Saffron Waldon: Ketton, 41; Totternhoe, 75

Bakewell
    St Mary's Hospital: Mandale, 45
    see also Ashford in the Water
Banbury: Wroxton, 79
    see also Broughton Castle
Bath
    15 to 20 Beauford Square: Monks Park, 17; Westwood Ground, 17
    Green Park Station: Combe Down, 15
    1 Henrietta Street: Monks Park, 17; Westwood Ground, 17
    Marks & Spencer: Monks Park, 17
    Pulteney Weir: Doulting, 27
    Stall Street Pump House: Monks Park, 17; Westwood Ground, 17
    Woolworth: Monks Park, 17
    York Street Pump House: Monks Park, 17
    see also Combe Down
Bedford
    Marlborough School: Ketton, 42
    School: Weldon, 78
    Swan Hotel: Totternhoe, 75
Belfast, St Anne's Cathedral: Ham Hill, 33; Portland, 55
Belvoir Castle, nr Grantham: Ancaster, 13
Bicester, Stratton Audley Manor: Guiting, 31
Birkenhead, Mersey Tunnel: Portland, 55
Birmingham Cathedral: Hopton Wood, 36; Orton Scar, 48
Blackburn Cathedral: Swaledale Fossil, 72
Bourne, *see* Grimsthorpe Castle
Bournemouth
    Crematorium: Purbeck Limestone, 65
    St Peter's Church: Westwood Ground, 20
Brackley, *see* Hunton-in-the-Hedges
Bradford Library: Portland, 55
Brentford Cathedral: Kentish Ragstone, 66; Purbeck Limestone, 66
Bridgewater, *see* North Petherton
Bridport, *see* Chideock
Brighton Pavilion: Combe Down, 16
Bristol
    Bishopsworth Church: Stowey, 71
    Clifton Cathedral: Purbeck-Portland, 68
    Clifton Down Station: Monks Park, 20; Westwood Ground, 20
    First Church of Christian Science: Monks Park, 17; Westwood Ground, 17

Bristol (continued)
    Lichfield Road: Stowey, 71
    Old Clifton Down Station: Stowey, 71
    Sun Alliance House: Monks Park, 17
Bristol Polytechnic, Clifton: Stowey, 71
Broad Campden: Happylands, 35
Broughton Castle, nr Banbury: Hornton, 37
Burghley House, Stamford: Ketton, 42
    Bottle Lodges: Ketton, 41
Burlescombe, nr Tiverton: Fenacre, 29

Calne, Chippenham Road: Daglingworth, 26
Cambridge
    Great St Mary's Church: Weldon, 77
    Westcott House, Jesus Lane: Ham Hill, 33
Cambridge University
    Downing College: Ketton, 41
    Fitzwilliam Museum: Portland, 50
    Gonville and Caius College: Ancaster, 13; Weldon, 77
    Jesus College: Weldon, 77
    King's College: Clipsham, 22; Weldon, 22
        Chapel: Ancaster, 13; Clipsham, 77; Ketton, 42; Weldon, 77
        Fellows' Building: Portland, 50
    Library: Hornton, 37
    Old University Library: Portland, 50
    Peterhouse, Old Court: Ancaster, 13
    St Catharine's College: Ketton, 41
    St John's College, Chapel: Ancaster, 13
    Senate House: Portland, 50
    Sidney Sussex College: Weldon, 77
    Trinity College: Ketton, 41; Weldon, 77
        Master's Lodge: Ham Hill, 33
    Trinity Hall: Ancaster, 13
    Wolfson College: Clipsham, 77; Weldon, 77
Canterbury Cathedral: Clipsham, 22
Cardiff, Grangetown and Lansdowne Primary School: Westwood Ground, 20
Carrick on Suir, bridge: Lecarrow, 44
Cattistoke, *see* Chantmarle
Chantmarle, nr Cattistoke: Ham Hill, 33
Chard, *see* Forde Abbey
Chatsworth, Derbyshire: Hopton Wood, 36
Cheltenham
    Cheltenham and Gloucester Building Society: Monks Park, 17
    Leckhampton Court: Monks Park, 17; Westwood Ground, 17
    Police Headquarters: Monks Park, 17
Chideock Church, nr Bridport: Ham Hill, 33
Cirencester
    Dollar Street: Westwood Ground, 20
    Gloucester Road: Daglingworth, 26
    Police Headquarters: Monks Park, 17
    Tesco Superstores: Guiting, 31
Clayworth, nr Gainsborough, St Peter's Church: Ancaster, 13
Clipsham: Guiting, 30
Combe Down, nr Bath, Prior Park College: Combe Down, 16

Corby, *see* Geddington
Corfe Castle: Purbeck Limestone, 60, 65
Cork, Parnell Bridge: Lecarrow, 44
Coventry
    Cathedral: Kilkenny Marble, 43
    Drapers Hall: Stoke Ground Base Bed, 19

Derby, *see* Kedleston Hall
Dorchester, *see* Sturminster Newton
Dover Castle: Purbeck Limestone, 65
Dublin: Portland, 50
    Christchurch Cathedral: Lecarrow, 44
    Garden of Remembrance: Ballinasloe, 14
    Russel Court Hotel: Ballinasloe, 14
    University, Belsied College: Lecarrow, 44
    University College: Ballinasloe, 14; Kilkenny Marble, 43
Dunstable
    Priory: Totternhoe, 75
    see also Eaton Bray
Durham Cathedral: Purbeck Marble, 60

Eaton Bray, nr Dunstable, Church: Totternhoe, 75
Ely
    Bishop's Palace: Weldon, 77
    Cathedral: Ancaster, 12; Clipsham, 22; Purbeck Marble, 60; Weldon, 78
    Tower Hospital: Ketton, 42
Exeter Cathedral: Purbeck Marble, 60

Fetcham, Leatherhead, Fetcham Park House: Westwood Ground, 20
Fleshford: Combe Down, 16
Forde Abbey, nr Chard: Ham Hill, 33
Frome, Bristol Water Works: Stowey, 71
Fulham, All Saints Church: Westwood Ground, 20

Gainsborough, *see* Clayworth
Galway Cathedral: Ballinasloe, 14; Kilkenny Marble, 43
Geddington, nr Corby, Queen Eleanor Cross: Weldon, 77
Gilling East: Hovingham, 38
Glasgow
    Burrel Museum: Portland, 57
    Clydesdale Bank HQ: Swaledale Fossil, 72
Glastonbury Abbey: Doulting, 27
Gloucester, *see* Todenham
Grantham
    National Westminster Bank: Ancaster, 13
    see also Belvoir Castle, Harlaxton Manor, Kesteven Hospital
Grimsthorpe Castle, nr Bourne: Ketton, 42
Guildford Cathedral: Doulting, 27

Hampton Court Palace: Clipsham, 22; Portland, 50; Stoke Ground Base Bed, 19
Harlaxton Manor, nr Grantham: Ancaster, 13
Hawton, nr Newark, Church: Ancaster, 13
Heathrow Airport, Terminal 2: Swaledale Fossil, 72
Helmsley: Hovingham, 38
    see also Rievaulx Abbey
Hitchin Priory: Totternhoe, 75
Hornton: Wroxton, 79
Horsham, Sun Alliance Building: Portland, 55
Hovingham: Hovingham, 38
Huish Episcopi, nr Taunton, Church: Ham Hill, 33
Hunslet, nr Leeds, Methodist Church: Hovingham, 38
Hunton-in-the-Hedges, nr Brackley: Guiting, 30

Isle Abbotts, nr Taunton, Church: Ham Hill, 33

Kedleston Hall, nr Derby: Hopton Wood, 36
Kendal
    Abbot Hall: Orton Scar, 48
    Kentmear House: Salterwath, 70
    Trustee Savings Bank: Orton Scar, 48
Kesteven Hospital, nr Grantham: Ancaster, 13
Kington: Nash Rocks, 46
Knole House, nr Sevenoaks: Kentish Ragstone, 39

Lancaster University, County College: Orton Scar, 48
Lancing College Chapel: Doulting, 27
Langton Matravers: Purbeck Limestone, 63
Leatherhead, *see* Fetcham
Leeds
    Interland Hotel: Swaledale Fossil, 72
    see also Hunslet
Leicester, Magnum Hotel: Salterwath, 70
Lincoln
    Cathedral: Purbeck Marble, 60
    Chapter House: Weldon, 78
    Gowt's Bridge, High St: Ancaster, 13
    Guildhall: Ancaster, 13
London
    AUEW Headquarters: Portland, 55
    All Saints Russian Orthodox Church, Ennismore Gardens: Doulting, 27
    Bank of England: Portland, 50
    Barclay's Bank, Gracechurch Street: Portland, 57
    Berkeley Hotel, Knightsbridge: Clipsham, 22
    Blackwall Tunnel: Portland, 55
    British Museum: Portland, 50
    Buckingham Palace: Clipsham, 22
    City Prison, Holloway: Kentish Ragstone, 39
    County Hall: Portland, 50
    Cutlers Gardens office complex: Portland, 52
    Devonshire House, Piccadilly: Portland, 55
    Geological Museum: Hopton Wood, 36
    Guildhall: Doulting, 27; Hopton Wood, 36; Portland, 27
    Houses of Parliament *see* (London) Palace of Westminster
    Intercontinental Hotel, Park Lane: Portland, 55
    Kenstead Hall: Portland, 55
    Lambeth Palace: Stoke Ground Base Bed, 19
    Palace of Westminster: Clipsham, 22; Ketton, 41
        Peers' Entrance: Clipsham, 25
        Westminster Hall: Combe Down, 16
    Polish Katyn Memorial, Gunnersbury Park Cemetery: Portland, 57
    Queen Anne's Mansions: Portland, 55
    Red Lion Pub, Whitehall: Portland, 52
    Royal Commonwealth Society: Ancaster, 12
    Royal Festival Hall: Portland, 55
    St Paul's Cathedral: Clipsham, 73; Portland, 50; Taynton, 73
    Somerset House: Portland, 50
    Tower Bridge: Portland, 57
    Tower Hill: Mandale, 45
    Tower of London: Purbeck Limestone, 65
        White Tower: Kentish Ragstone, 39
    Victoria and Albert Museum: Hopton Wood, 36
    Well Hall Police Station, Eltham: Taynton, 73
    Westminster Abbey: Purbeck Marble, 60
    Woolworth, Cheapside: Portland, 55
London Colney, *see* Salisbury Hall
Longdon, nr Tewkesbury, B4211: Daglingworth, 26

Maidstone
    Bridge: Kentish Ragstone, 39
    Crown Court: Portland, 55
Manchester
    Cathedral: Swaledale Fossil, 72
    Magnum Hotel: Salterwath, 70
    Town Hall: Hopton Wood, 36
Menai Strait, N Wales
    Britannia Bridge: Salterwath, 70
    Menai Bridge: Penmon, 49

Newark
    Church and castle: Ancaster, 13
    see also Hawton
North Petherton, nr Bridgewater, Church: Ham Hill, 33
Northampton
    C & A Modes, Market Square: Weldon, 78
    Welsh House, Market Square: Hornton, 37, 78; Weldon, 78
Norwich
    Cathedral: Clipsham, 22; Weldon, 78
    St Peter Mancroft Church: Ancaster, 12
Nottingham, see Wollaton Hall

Old Soar Manor, nr Plaxtol: Kentish Ragstone, 39
Oxford
    Carfax: Clipsham, 22
    Magdalen Bridge: Taynton, 73
    Playhouse, Beaumont Street: Clipsham, 22
    Royal Oxford Hotel: Guiting, 30
Oxford University
    All Souls College: Clipsham, 22
    Ashmolean Museum: Portland, 50
    Balliol College: Guiting, 30
    Bodleian Library: Taynton, 73
    Bodleian (Old Schools) quadrangle: Clipsham, 22
    Brasenose College, New Quadrangle: Clipsham, 22
    Christ Church
        Library: Clipsham, 22; Portland, 50
        The Tom Tower: Clipsham, 22
        West front: Clipsham, 22
    Examination Schools: Clipsham, 22
    Green College: Ancaster, 13; Clipsham, 25; Daglingworth, 26; Purbeck, 60; Weldon, 78
    Old Observatory: Weldon, 78
    Inorganic Chemistry Laboratory: Clipsham, 22, 25
    Lincoln College: Clipsham, 22
        Grove buildings: Clipsham, 22
    Magdalen College, Bursar's Lodge: Guiting, 30
    Manchester College, Arlosh Hall: Clipsham, 22
    Merton College
        St Alban's Quadrangle: Weldon, 78
        Warden's House: Weldon, 78
    Nuffield College: Clipsham, 22, 25
    Old Ashmolean Building: Clipsham, 22
    Oriel College, Rhodes Building: Weldon, 78
    St John's College: Taynton, 73
    Sheldonian Theatre: Clipsham, 22
    Somerville College, Chapel: Clipsham, 22
    The Queen's College: Portland, 50

Peterborough
    Cathedral: Clipsham, 22; Totternhoe, 75
    New shopping development: Clipsham, 23
Plaxtol, see Old Soar Manor
Portland, Rufus Castle: Portland, 50
Preston, County Hall: Orton Scar, 48
Prinknash Abbey, Chapel: Guiting, 30

Radstock, Bristol Water Works: Stowey, 71
Ratley, nr Warmington: Hornton, 37
    Church: Hornton, 37
Reading, 179, 181–183, 185 Kings Road: Monks Park, 18; Westwood Ground, 18
Reculver: Purbeck Limestone, 61
Rievaulx Abbey, nr Helmsley, Ionic Temple: Ancaster, 13
Ripon Cathedral: Clipsham, 22
Ripple, nr Tewkesbury, Road and Railway Bridge: Daglingworth, 26
Rochester
    Castle: Kentish Ragstone, 39
    Cathedral: Weldon, 78
Roscommon, Bank of Ireland: Lecarrow, 44

Saffron Waldon, see Audley End House
St Albans Cathedral: Totternhoe, 75
Salisbury Cathedral: Clipsham, 22; Purbeck Marble, 60
Salisbury Hall, nr London Colney: Totternhoe, 75
Sevenoaks, see Knole House
Sheffield, Grosvenor House Hotel: Portland, 55
Shepton Mallet, new town centre: Doulting, 27; Purbeck Limestone, 27, 60
Sherborne, Hospital of St John: Ham Hill, 33
Shipston on Stour, Honington Bridge: Guiting, 32
Shutford, Old Bake House: Wroxton, 79
Slough Library: Portland, 55
Somerton, Church, Hext Almshouses, Town Hall: Ham Hill, 33
Stamford
    Hospital: Ketton, 41
    see also Burghley House
Stoke-on-Trent Library: Portland, 55
Stratford-on-Avon
    Hilton Hotel: Hornton, 37
    National Farmers' Union and Avon Mutual Insurance Co building: Monks Park, 18; Westwood Ground, 18
Sturminster Newton, Dorchester: Guiting, 30
Swindon, Princess Margaret Hospital: Daglingworth, 26

Tattershall Castle: Ancaster, 12
Taunton
    County offices: Doulting, 27
    4–5 Fore Street: Ham Hill, 33
    Westminster Bank, Fore Street: Monks Park, 17; Westwood Ground, 17
    see also Huish Episcopi, Isle Abbotts
Taynton: Taynton, 73
Tewkesbury
    Tudor Gatehouse, nr Abbey: Guiting, 30
    see also Longdon, Ripple
Tiverton, see Burlescombe
Todenham, Gloucester, Church: Westwood Ground, 20

Ulverston Diversion, New Church Lane: Mandale, 45

Warmington, see Ratley
Warndon, nr Worcester: Doulting, 27
Warwick, Warwickshire County Council offices: Hornton, 37
Wells Cathedral: Doulting, 27; Kilkenny Marble, 43
West Coker Manor, nr Yeovil: Ham Hill, 33
Weybridge, Hamstone House: Ham Hill, 33
Winchester
    Law Courts: Clipsham, 22
    School: Purbeck-Portland, 68
Windsor Castle: Clipsham, 22; Stoke Ground Base Bed, 19; Taynton, 73

Witney
  Halifax Building Society: Guiting, 31; Westwood Ground, 20
  National Westminster Bank: Guiting, 31
Woburn Abbey: Ketton, 41; Totternhoe, 74
Wollaton Hall, nr Nottingham: Ancaster, 13
Woodstock, Town Hall: Taynton, 73
Worcester
  Library: Hornton, 37
  see also Warndon
Worthing, Gateway Building Society: Portland, 55
Wroxton College: Hornton, 37

Yeovil
  Montacute House: Ham Hill, 33
  St John's Church: Ham Hill, 33
  see also West Coker Manor
York
  Cathedral: Clipsham, 22
  Garrowby Hall: Ancaster, 13